33rd Annual CSP Workshop: Recent Developments in Computer Simulation Studies in Condensed Matter Physics (CSP 2020)

Journal of Physics: Conference Series Volume 2241

Online
17-21 February 2021

ISBN: 978-1-7138-5496-8
ISSN: 1742-6588

Printed from e-media with permission by:

Curran Associates, Inc.
57 Morehouse Lane
Red Hook, NY 12571

CURRAN ASSOCIATES INC. proceedings.com

Some format issues inherent in the e-media version may also appear in this print version.

Institute of Physics
Dirac House, Temple Back
Bristol BS1 6BE UK

Phone: 44 1 17 929 7481
Fax: 44 1 17 920 0979

techtracking@iop.org

Additional copies of this publication are available from:

Curran Associates, Inc.
57 Morehouse Lane
Red Hook, NY 12571 USA
Phone: 845-758-0400
Fax: 845-758-2633
Email: curran@proceedings.com
Web: www.proceedings.com

TABLE OF CONTENTS

Journal of Physics: Conference Series **2241** (2022) 011001 doi:10.1088/1742-6596/2241/1/011001

Preface

Thirty four years ago, because of the dramatic increase in the power and utility of computer simulations, The University of Georgia formed the first institutional unit devoted to the application of simulations in research and teaching: The Center for Simulational Physics. Then, as the international simulations community expanded further, we sensed the need for a meeting place for both experienced simulators and newcomers to discuss inventive algorithms and recent results in an environment that promoted lively discussion. As a consequence, the Center for Simulational Physics established an annual workshop series on *Recent Developments in Computer Simulation Studies in Condensed Matter Physics*. This year's highly interactive workshop was the 33rd in the series marking our efforts to promote high quality research in simulational physics. The continued interest shown by the scientific community amply demonstrates the useful purpose that these meetings have served.

The latest workshop was held at The University of Georgia from February 17-21, 2020. These Proceedings provide a "status report" on a number of important topics. This on-line "volume" is published with the goal of timely dissemination of the material to a wider audience.

These Proceedings contain both invited papers and contributed presentations on problems in both classical and quantum condensed matter physics. The Workshop topics, as usual, ranged from hard and soft condensed matter to biologically inspired problems, and quantum computing. We hope that readers will benefit from specialized results as well as profit from exposure to new algorithms, methods of analysis, and conceptual developments.

D. P. Landau
M. Bachmann
S. P. Lewis
H. B. Schüttler

Please see below the full list of committee's names and affiliations which shall be published as part of the preface with their contact details and institutional email address.

1. Prof. Heinz-Bernd Schuttler
University of Georgia
220 Cedar St.
Athens, GA 30602
Email: hbs@uga.edu

2. Prof. Michael Bachmann
University of Georgia
220 Cedar St.
Athens, GA 30602
Email: mbach01@smsyslab.org

3. Prof. Steven Lewis
University of Georgia
220 Cedar St.
Athens, GA 30602
Email: splewis@uga.edu

4. Prof. David Landau
University of Georgia
220 Cedar St.
Athens, GA 30602
Email: dlandau@uga.edu

Peer Review Statement

All papers published in this volume have been reviewed through processes administered by the Editors. Reviews were conducted by expert referees to the professional and scientific standards expected of a proceedings journal published by IOP Publishing.

- **Type of peer review:** Single Anonymous
- **Conference submission management system:** Morressier
- **Number of submissions received:** 8
- **Number of submissions sent for review:** 8
- **Number of submissions accepted:** 8
- **Acceptance Rate (Submissions Accepted / Submissions Received × 100):** 100
- **Average number of reviews per paper:** 1
- **Total number of reviewers involved:** 4
- **Contact person for queries:**
  Name: David P. Landau
  Email: dlandau@uga.edu
  Affiliation: Center for Simulational Physics

Quantum Monte Carlo Study of Weyl Superconductivity

Peter Rosenberg(1) **and Efstratios Manousakis**(1,2)

(1)National High Magnetic Field Laboratory, Florida State University, Tallahassee, FL 32306-4350, USA

(2)Department of Physics, University of Athens, Panepistimioupolis, Zografos, 157 84 Athens, Greece

E-mail: manousakis@magnet.fsu.edu

Abstract. Using auxiliary-field quantum Monte Carlo we study superconductivity in a simple extension of the two-dimensional negative-U Hubbard model which hosts Weyl nodes near the Fermi surface. We discuss the model and our method as applied to the study of superconductivity in this system. We discuss the nature of pairing between Weyl quasiparticles which carry opposite spin and opposite topological charge leading to a spin-singlet pairing amplitude and to a mixture of pseudo-spin singlet and pseudo-spin triplet pairing.

1. Introduction

There is a long history of applications of topological concepts to physical systems, with the gauge-field theories of nuclear and particle physics being perhaps the most prominent examples. Recently, the notion of topology has become a central focus of condensed matter physics. This renewed interest has been largely motivated by the discovery of the so-called topological insulators, followed by the more recent discovery of materials that host Weyl quasiparticles and transition to a superconducting phase at low temperatures [1, 2, 3, 4, 5, 6]. The origin of superconductivity in Weyl materials is not yet well-understood, particularly beyond the level of mean-field theory, which makes a quantitative many-body description of the pairing behavior in these systems an important goal. Such a description will help resolve many of the questions surrounding the unique properties of these materials, for instance, whether or not pairing between quasiparticles of opposite chirality (topological charge) is responsible for the emergence of the observed superconductivity. A more complete understanding of the properties of these materials will serve as a foundation for many exciting potential applications across a variety of fields including quantum computation and information.

Here, we report the results from a numerically-exact auxiliary-field quantum Monte Carlo (AFQMC) study of a 2D Weyl model. The model is obtained from a minimal modification of the well-known negative-U Hubbard model that converts it into one of the simplest models to exhibit Weyl character. We center our investigation on pairing between the Weyl quasiparticles that emerge in the system, but our results provide a thorough quantitative description of the ground-state properties of the system, including the charge and pseudo-spin behavior. This work establishes the capability of AFQMC to treat interacting topological systems, and provides a reliable approach for future quantitative investigations of topological superconductivity.

A large volume of results from this work has been published elsewhere [7]. In the present paper, we focus on some of the technical aspects of the quantum MC method as applied to this model and, in addition, we highlight and summarize the main results.

2. Our Model

The tight-binding component of our model is constructed by stacking SSH chains which are coupled by means of a diagonal hopping t_d, as follows,

$$\hat{H}_0 = -\sum_{n,m,\sigma}\left[vc^{(A)\dagger}_{n,m\,\sigma}c^{(B)}_{n,m\,\sigma} + wc^{(A)\dagger}_{n,m\,\sigma}c^{(B)}_{n-1,m\,\sigma} + t_d(c^{(B)\dagger}_{n,m\,\sigma}c^{(A)}_{n,m\pm1,\sigma} + c^{(B)\dagger}_{n,m\,\sigma}c^{(A)}_{n+1,m\pm1\,\sigma})\right] + \text{h.c.} \quad (1)$$

where the operator $c^{(A)\dagger}_{n,m\,\sigma}$ creates an electron of spin σ on the A site of the unit cell at position $\mathbf{R} = na\,\hat{x} + mb\,\hat{y}$. Each unit cell is composed of an A site and a B site. The intra- and inter-unit-cell hopping strengths in the $\hat{x}$-direction are controlled by the parameters v and w, respectively.

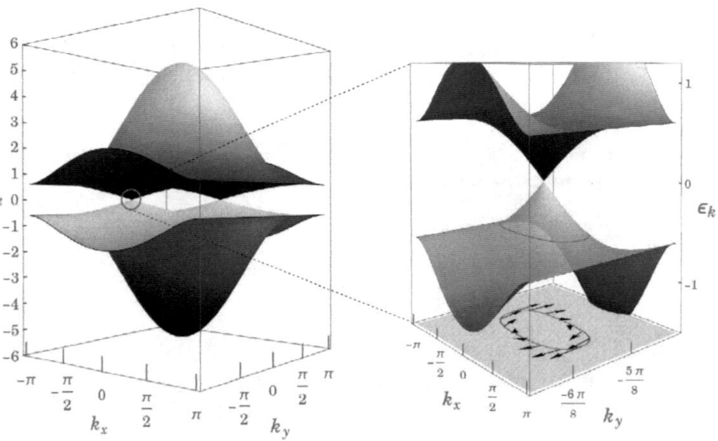

Figure 1. Band structure of our model of a 2D Weyl system. On the right we highlight one of the nodal points, whose chirality is reflected by the vorticity of the Berry potential. We plot the Berry potential (black arrows) around the Fermi surface (red contour) for a system with density, $n \sim 0.95$.

In momentum space the Hamiltonian can be written,

$$\hat{H}_0 = \sum_{\mathbf{k}\sigma} \mathbf{c}^{\dagger}_{\mathbf{k}\sigma}\mathcal{H}(\mathbf{k})\mathbf{c}_{\mathbf{k}\sigma}\,, \quad (2)$$

with the vector $\mathbf{c}^{\dagger}_{\mathbf{k}\sigma} = (c^{(A)\dagger}_{\mathbf{k}\sigma}, c^{(B)\dagger}_{\mathbf{k}\sigma})$, and the matrix $\mathcal{H}(\mathbf{k}) = -\mathbf{h}(\mathbf{k})\cdot\boldsymbol{\sigma}$. This has the form of a Weyl Hamiltonian, where $\boldsymbol{\sigma} = (\sigma_x, \sigma_y, \sigma_z)$ is a vector of the Pauli matrices in the sublattice (or pseudo-spin) basis, and

$$\mathbf{h}(\mathbf{k}) = \begin{pmatrix} v_1 + w_1\cos(k_x a) \\ w_1\sin(k_x a) \\ 0 \end{pmatrix}, \quad (3)$$

with $v_1 = v + 2t_d\cos(k_y b)$, and $w_1 = w + 2t_d\cos(k_y b)$. We plot the band structure in Fig. 1. For $v + w < 4t_d$ there are two Weyl nodes at $\mathbf{k} = (0, \pm k_N)$, with $k_N = \cos^{-1}(-(v+w)/4t_d)$. These nodes are protected by the combined inversion (A $\to$ B, $\mathbf{k} \to -\mathbf{k}$) and time-reversal symmetry of the tight-binding Hamiltonian. For a more detailed characterization of the topological features of the model see Ref.[7].

The Hamiltonian in this $\mathbf{k} \cdot \boldsymbol{\sigma}$ notation is a 2×2 matrix, which can be diagonalized in the pseudo-helicity basis given in Ref.[7]. This introduces a new set of creation operators, $\chi^{(\pm)\dagger}_{\mathbf{k}\sigma}$ in terms of which the tight-binding Hamiltonian takes the form:

$$\hat{H}_0 = \sum_{\mathbf{k},\sigma,\alpha=\pm} \epsilon^{(\alpha)}_{\mathbf{k}} \chi^{(\alpha)\dagger}_{\mathbf{k}\sigma} \chi^{(\alpha)}_{\mathbf{k}\sigma}, \qquad \epsilon^{(\pm)}_{\mathbf{k}} = \pm\sqrt{(h^x_{\mathbf{k}})^2 + (h^y_{\mathbf{k}})^2}\,. \tag{4}$$

where $h^x_{\mathbf{k}}$ and $h^y_{\mathbf{k}}$ are components of the vector $\mathbf{h}(\mathbf{k})$.

We introduce interactions into the system via an on-site attractive Hubbard term,

$$\hat{H}_{\mathrm{I}} = \sum_{i,\alpha} U n^{\alpha}_{i\uparrow} n^{\alpha}_{i\downarrow}, \tag{5}$$

where i labels the unit cell, $\alpha = \mathrm{A}, \mathrm{B}$, and the interaction parameter $U < 0$. Our full Hamiltonian becomes $\hat{H} = \hat{H}_0 + \hat{H}_{\mathrm{I}}$.

3. Application of the AFQMC method

We treat our model with the auxiliary-field quantum Monte Carlo technique [8, 9, 10]. The method, which is based on the principle of imaginary-time projection from an initial trial wavefunction, has demonstrated considerable accuracy in applications to various problems (see Ref. [7] and references therein). In sign-problem-free cases, such as the attractive spin-unpolarized Hubbard model, or the Weyl model we consider in this work, AFQMC calculations are numerically exact, by which we mean there are no unconstrained approximations, and arbitrary accuracy can be achieved with a sufficient number of Monte Carlo samples. The technique can be used to calculate an array of ground state properties and provide a reliable quantitative many-body description of the pairing, spin, and charge behaviors of strongly correlated systems. An outline of the method is provided in Ref. [7] and in the present paper we only discuss its application to our problem.

We begin with a comparison of AFQMC results to those from exact diagonalization for small-size lattices. In Fig. 2 we compare the results of our exact diagonalization with AFQMC results for the bond-density and s-wave pair-pair correlation functions. The operator $\Delta^{\dagger}_i = c^{\dagger}_{i\uparrow} c^{\dagger}_{i\downarrow}$ creates a singlet pair at lattice site i, and the operator $\rho^x_\ell = c^{\dagger}_{\mathbf{r}_\ell} c_{\mathbf{r}_\ell + \hat{x}} + \mathrm{h.c.}$, measures the density on the ℓ-th bond along the x-direction. The system is a 2×3 unit-cell lattice with $v = 0.2$, $w = 1.0$, $t_d = 0.9$, $U = -8.0$ and $N = 8$ electrons. The results in Fig. 2(a) and Fig. 2(b) demonstrate that the implementation correctly reproduces the results of the exact diagonalization. The results in Fig. 2(c) demonstrate that the value of the projection time, β, used in order to produce the correct ground state from the chosen initial state is sufficiently large.

4. Highlight of our results

Figure 3 shows the calculated pairing amplitudes $\psi^{\alpha\beta}_{\mathbf{k}}$ as obtained from the pairing matrix defined from the matrix elements of the pairing operators defined as[7]

$$\Delta^{\alpha\beta}_{\mathbf{k}} = \frac{1}{\sqrt{2}} c^{(\alpha)\dagger}_{\mathbf{k}\uparrow} c^{(\beta)\dagger}_{-\mathbf{k}\downarrow} - c^{(\alpha)\dagger}_{\mathbf{k}\downarrow} c^{(\beta)\dagger}_{-\mathbf{k}\uparrow} \tag{6}$$

where α and β take values A and B according to their sublattice (pseudo-spin) character. Notice that $\psi^{\mathrm{AA}}_{\mathbf{k}}$ and $\psi^{\mathrm{BB}}_{\mathbf{k}}$ (left panel of Fig. 3) peak near the most "curved" parts of the Fermi surface

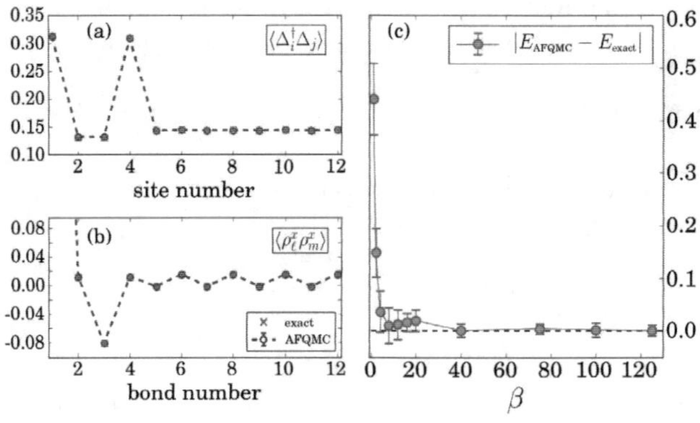

Figure 2. (a): Comparison of the pair correlation function with exact diagonalization results. (b) Comparison of the bond-density correlation function with exact diagonalization results. (c) Convergence of the ground state energy calculated with AFQMC as a function of projection time, β. We show the magnitude of the difference between the AFQMC and exact results (in units of v^{-1}).

where the density of states is large. In addition, we have separated the imaginary part of the pseudo-spin singlet (the real part is zero) and the real part of pseudo-spin triplet (its imaginary part is zero) component of the pairing amplitude $\psi_{\mathbf{k}}^{AB}$, which are plotted in the middle panel of Fig. 3. On the right panel of Fig. 3 we give the calculated pseudo-spin distribution inside the Brillouin zone. Notice the non-trivial nodal structure of the components of the pairing amplitude, especially of the pseudo-spin singlet (bottom-middle) part of Fig. 3. Generally speaking the pairing amplitude follows the behavior of the total pseudo-spin value of each pair of quasiparticles of opposite $\mathbf{k}$ as it varies along the Fermi surface. The pseudo-spin vectors of most pairs with momenta $\mathbf{k}$ and $-\mathbf{k}$ are almost parallel to each other, and this explains the dominance of the pseudo-spin triplet. On the other hand, the pseudo-spin singlet pairing amplitude (bottom-middle plot) peaks in the region of the Fermi surface where the two pseudo-spin vectors of the pair with momenta $\mathbf{k}$ and $-\mathbf{k}$ have the largest anti-parallel components, and it has a node where the vectors are parallel.

5. Conclusions

Using the cutting-edge AFQMC method we have studied the nature of superconductivity and the pairing state of Weyl quasiparticles that emerge in a simply modified [7] negative-U Hubbard model. We observe pairing between quasiparticles that carry opposite topological charge, resulting in a topological-charge-neutral pairing amplitude composed of real-spin singlets that are a mixture of pseudo-spin singlet and pseudo-spin triplet. The structure of the pairing amplitude is strongly affected by the dipole-like pseudo-spin distribution, which is generated by the pair of Weyl nodes in the system. Our results provide a quantitative description of the rich pairing behavior of this interacting topological system.

The interplay of topology and interactions in many-body systems is a challenging problem of considerable importance to modern condensed matter physics. This work demonstrates the power of the AFQMC method for studying strongly-interacting topologically non-trivial systems, and is an important step towards a more complete understanding of pairing and strong correlations in the context of topological electronic structure, which lays the foundation for future studies of topological superconductivity.

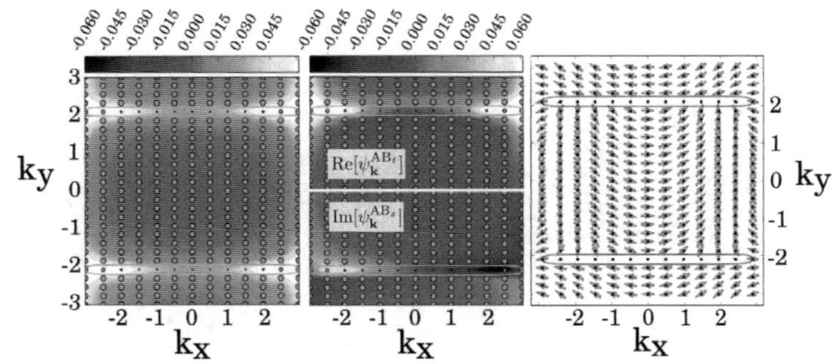

Figure 3. Left panel: Intensity plot of $\psi_{\mathbf{k}}^{AA}$ or $\psi_{\mathbf{k}}^{BB}$. Middle: An intensity plot of the real (imaginary) part of the triplet (singlet) component of $\psi_{\mathbf{k}}^{AB}$ is shown in the top (bottom) middle panel. Right: The pseudo-spin distribution inside the Brillouin zone. The system is a 13×27 unit-cell lattice with $v = 0.7$, $w = 1.1$, $t_d = 0.9$, $U = -1.2$, and $N = 658$ electrons.

6. Acknowledgments

This work was supported in part by the U.S. National High Magnetic Field Laboratory, which is funded by NSF DMR-1157490 and the State of Florida.

References

[1] Kang D, Zhou Y, Yi W, Yang C, Guo J, Shi Y, Zhang S, Wang Z, Zhang C, Jiang S, Li A, Yang K, Wu Q, Zhang G, Sun L and Zhao Z 2015 *Nature Communications* **6** 7804 URL `https://doi.org/10.1038/ncomms8804`

[2] Pan X C, Chen X, Liu H, Feng Y, Wei Z, Zhou Y, Chi Z, Pi L, Yen F, Song F, Wan X, Yang Z, Wang B, Wang G and Zhang Y 2015 *Nature Communications* **6** 7805 URL `https://doi.org/10.1038/ncomms8805`

[3] Qi Y, Naumov P G, Ali M N, Rajamathi C R, Schnelle W, Barkalov O, Hanfland M, Wu S C, Shekhar C, Sun Y, Süß V, Schmidt M, Schwarz U, Pippel E, Werner P, Hillebrand R, Förster T, Kampert E, Parkin S, Cava R J, Felser C, Yan B and Medvedev S A 2016 *Nature Communications* **7** 11038 article URL `https://doi.org/10.1038/ncomms11038`

[4] Guguchia Z, von Rohr F, Shermadini Z, Lee A T, Banerjee S, Wieteska A R, Marianetti C A, Frandsen B A, Luetkens H, Gong Z, Cheung S C, Baines C, Shengelaya A, Taniashvili G, Pasupathy A N, Morenzoni E, Billinge S J L, Amato A, Cava R J, Khasanov R and Uemura Y J 2017 *Nature Communications* **8** 1082 URL `https://doi.org/10.1038/s41467-017-01066-6`

[5] Rhodes D, Schönemann R, Aryal N, Zhou Q, Zhang Q R, Kampert E, Chiu Y C, Lai Y, Shimura Y, McCandless G T, Chan J Y, Paley D W, Lee J, Finke A D, Ruff J P C, Das S, Manousakis E and Balicas L 2017 *Phys. Rev. B* **96**(16) 165134 URL `https://link.aps.org/doi/10.1103/PhysRevB.96.165134`

[6] Rhodes D, Yuan N F, Jung Y, Antony A, Wang H, Kim B, Chiu Y c, Taniguchi T, Watanabe K, Barmak K, Balicas L, Dean C R, Qian X, Fu L, Pasupathy A N and Hone J 2019 *arXiv e-prints* (*Preprint* `1905.06508`)

[7] Rosenberg P, Aryal N and Manousakis E 2019 *Phys. Rev. B* **100**(10) 104522 URL `https://link.aps.org/doi/10.1103/PhysRevB.100.104522`

[8] Sugiyama G and Koonin S 1986 *Annals of Physics* **168** 1 – 26 ISSN 0003-4916 URL `http://www.sciencedirect.com/science/article/pii/0003491686901077`

[9] Zhang S 2013 *Emergent Phenomena in Correlated Matter: Modeling and Simulation* vol 3 ed Pavarini E, Koch E and Schollwöck U (Verlag des Forschungszentrum Jülich, Jülich, Germany)

[10] Assaad F F 2002 Quantum monte carlo methods on lattices: The determinantal approach *Quantum Simulations of Complex Many-Body Systems: From Theory to Algorithms* vol 10 ed Grotendorst J, Marx D and Muramatsu A (NIC, Jülich, Germany) pp 99–156

Journal of Physics: Conference Series **2241** (2022) 012002 doi:10.1088/1742-6596/2241/1/012002

Scaling Theory for Driven Polymer Translocation through a Double Nanopore

Aniket Bhattacharya

Department of Physics, University of Central Florida, Orlando, USA

E-mail: `Aniket.Bhattacharya@ucf.edu`

Abstract. We study DNA translocation through a double nanopore system subject to a net bias using Brownian dynamics simulation on a model system. We consider the limit $d_{LR} << L$, where d_{LR} is the distance between the pores and $L = N\sigma$ is the contour length of the chain consisting of N monomers of diameter σ. In this limit, we generalize a scaling ansatz for the mean first passage time, originally proposed for the driven translocation through a single nanopore, for the double nanopore system and demonstrate its validity using simulation data. The simulation data enables us to extract the pore friction as a function of the chain stiffness. The method can be used to determine the mean first passage time $\langle\tau\rangle$ for longer chains difficult to extract from BD simulation.

1. Introduction

Polymer translocation through nanopore has remained an active field of study for almost two decades [1]. The original experiment where a single stranded DNA (ssDNA) was driven through an α-Hemolysin protein nanopore demonstrated that the method can possibly be used to determine DNA sequences without amplification and therefore, will be free from statistical error [2]. Since then the single nanopore (SNP) translocation has been carried out in synthetic nanopores. The method recently has been extended to a double nanopore (DNP) system which offers more promises [3]-[6]. In a DNP system, the presence of two forces at each pore offers better control, furthermore, a segment of the translocating polymer can be scanned multiple times by reversing the bias periodically. Numerical simulation studies have played a prominent role in verifying various theories of SNP polymer translocation. These studies showed that for the SNP translocation the finite chain length effect is strong due to nanopore-polymer interaction, and a scaling theory has been proposed to account for the discrepancy of the numerical results with theoretical predictions [7, 8, 9]. One of the advantages of the scaling theory is that one can extrapolate the scaling results for a longer chain. The original scaling ansatz for the SNP translocation was proposed for a fully flexible chain. Here we describe how this scaling ansatz for the SNP translocation can be generalized for a semiflexible chain translocating through a DNP system. We validate our hypothesis by carrying out Brownian Dynamics (BD) simulation on a bead-spring model of a polymer chain [10].

2. The Double-Nanopore Model and BD simulation

We have implemented the BD simulation scheme in two dimensions (2D) on a bead-spring model of Grest and Kremer [10]. The schematics are shown in Fig. 1 for $L = 21\sigma$, and $d_{LR} = 5\sigma$,

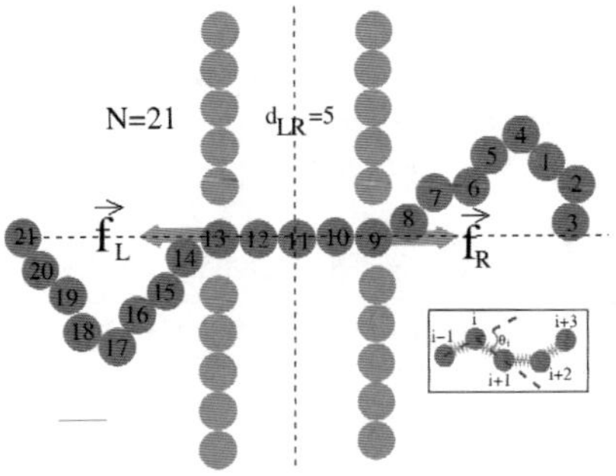

Figure 1. Schematics of the model DNP system shown for a chain length $L = 21\sigma$ and $d_{LR}/\sigma = 5$ with monomer beads (red) labeled from 1 to 21. The repulsive wall consists of monomers (teal) of the same diameter as that of the chain monomers. The forces $\vec{f}_L$ and $\vec{f}_R$ are applied at the center of each pore in opposite directions. The inset shows the detailed of the bead-spring model (Eqns. 1-3).

although the actual simulation runs are carried out for various chain lengths $L = 96\sigma - 384\sigma$ and for $d_{LR} = 16\sigma$. We start with an initial configuration where the polymer segment is co-captured in both the pores (Fig. 1). The origin of the coordinate system is chosen at the symmetric location equidistant from the two pores. Initially, a straight polymer chain is placed symmetrically with respect to the origin so that the equal straight segments lie at the left side of the left pore and at the right of the right pore so that at $t = 0$ the beads with indices 9-13 are located in between the right and the left pore, beads with indices $1 - 8$ and $14 - 21$ are located at the left and the right side of the left and the right pore respectively. The initial straight chain is then equilibrated with BD simulation time approximately equal to five times the Rouse relaxation time $\tau_{eq} \propto N^{1+2\nu}$ ($\nu = 0.75$ is the Flory exponent in 2D) keeping the beads located inside each pore clamped. To model the polymer chain, the immobile wall which contains the nanopore, we have used Lennard-Jones (LJ) and anharmonic spring potentials, and a three-body bond-bending potential as in Eqns. 1-3, described in detail in our recent publication [11]. The excluded volume interaction between any two monomers is given by a short-range Lennard-Jones (LJ) potential.

$$\begin{aligned} U_{\mathrm{LJ}}(r) &= 4\epsilon\left[\left(\frac{\sigma}{r}\right)^{12} - \left(\frac{\sigma}{r}\right)^{6}\right] + \epsilon, \text{ for } r \le 2^{1/6}\sigma; \\ &= 0, \text{ for } r > 2^{1/6}\sigma. \end{aligned} \tag{1}$$

Here, ϵ is the strength of the potential. The connectivity between neighboring monomers is modeled as a Finite Extension Nonlinear Elastic (FENE) spring with

$$U_{\mathrm{FENE}}(r) = -\frac{1}{2}kR_0^2 \ln\left(1 - r^2/R_0^2\right). \tag{2}$$

where r is the distance between the consecutive monomers, k is the spring constant and R_0 is the maximum allowed separation between connected monomers. The chain stiffness κ is introduced by adding an angle-dependent three-body interaction term between successive bonds as

$$U_{\mathrm{bend}}(\theta_i) = \kappa\left(1 - \cos\theta_i\right). \tag{3}$$

Here θ_i is the angle between the bond vectors $\vec{b}_{i-1} = \vec{r}_i - \vec{r}_{i-1}$ and $\vec{b}_i = \vec{r}_{i+1} - \vec{r}_i$, respectively (Fig. 1). The parameter $\kappa = \frac{1}{2}k_BT\ell_p/\sigma$, where ℓ_p is the persistence length of the chain, k_B and T are the Boltzmann constant and the temperature respectively. The values of the parameters and the BD scheme are the same as described in [11].

After the polymer chain is equilibrated with beads inside the left and right pores at the clamped position, the chain is allowed to translocate with biases applied at the left and the right pores as shown in Fig. 1. We consider various chain sizes $N = 96, 128, 192, 256, 320$, and

384, two different biases $|\vec{f}_R| - |\vec{f}_L| = 2$ and 3, and two different chain stiffness values $\kappa = 4$ and 8 respectively. The units of length, energy, and force are σ, ϵ, and ϵ/σ respectively. The simulation results are averaged over 2000 initial conditions. We have checked in few cases that a choice of the initial configuration of the chain as an off-lattice self-avoiding random walk does not affect the final results.

3. Results: Translocation with a net bias

As in an actual experiment, we apply a net bias $\Delta\vec{f}_{LR} = \vec{f}_L + \vec{f}_R \neq 0$ to the DNP system so that the DNA moves in the direction of the net bias. The presence of two forces at each pore can provide a feedback mechanism to control the movement of the translocating chain [4]. We define $\Delta f_{LR} = \left|\Delta\vec{f}_{LR}\right|$, and choose $\left|\vec{f}_R\right| > \left|\vec{f}_L\right|$, so that the direction of translocation is from the left pore to the right pore. First, we study the chain length dependence of the mean first passage

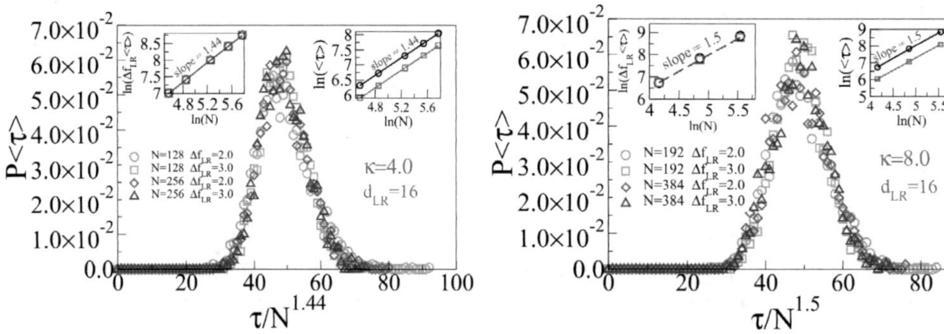

Figure 2. Histograms of the MFPT as a function of N^α for two different net bias $\Delta F_{LR} = 2$ and 3 and two chain lengths $N = 128$ and 256 respectively which show data collapse for (a) for $\kappa = 4.0$ and $\alpha = 1.44$, and (b) for $\kappa = 8.0$ and $\alpha = 1.50$. In each figure teal circles and orange squares correspond to $N = 128$ and $\Delta F_{LR} = 2$ and 3, and green diamonds and blue circles correspond to $N = 256$ and $\Delta F_{LR} = 2$ and 3. The insets at the right of each figure show log-log plots for $\langle\tau\rangle \sim N$ for $\Delta F_{LR} = 2$ (black circles) and 3 (red squares). The straight lines in each case is a linear fit with slope (a) $\alpha = 1.44 \pm 0.03$ and $\alpha = 1.50 \pm 0.03$. The insets at the left in each figure are the corresponding log-log plots of $\Delta F_{LR}\langle\tau\rangle \sim N$.

time (MFPT) for two values of the stiffness constant $\kappa = 4.0$ and 8.0 respectively (Fig. 2). BD simulation results show a power law for the MFPT $\langle\tau\rangle \sim N^\alpha$, where the translocation exponent α is a function of the chain stiffness κ ($\alpha = 1.44$ and 1.50 for $\kappa = 4.0$ and 8.0 respectively). This is evident from (i) the data collapse of the histogram of the MFPT and the (ii) two insets of Fig. 2. By combining these results with the fact that $\langle\tau\rangle \sim |\Delta f_{LR}|^{-1}$ we obtain

$$\langle\tau\rangle = C|\Delta f_{LR}|^{-1}N^\alpha. \tag{4}$$

Here C is a constant that can depend on the chain stiffness κ. The inset at the left of Fig. 2(a)-(b) convincingly shows the data collapse for two chain lengths $N = 128$ and 256 for two values of $\Delta f_{LR} = 2$ and 4 respectively, verifying Eqn. 4. This power law scaling of Eqn. 4 with the value of the effective translocation exponent $\alpha \approx 1.4 - 1.5$ for chain lengths $N \sim 100 - 500$ is the same as observed in a biased SNP [12, 13, 14]. This is discussed in detail below.

The scaling ansatz for the MFPT of a fully flexible chain ($\kappa = 0$) in the context of a SNP translocation is given by [7, 8, 9]

$$\langle\tau\rangle \sim \left[AN^{1+\nu} + \eta_{pore}N\right]|\Delta F_{LR}|^{-1} \sim N^\alpha. \tag{5}$$

Here A is constant that depends on κ, $\nu = 0.75$ and 0.5888 is the Flory exponent in 2D and three dimensions (3D), respectively, and η_{pore} is the pore friction. Eqn. 5 for a self-avoiding fully flexible chain holds for small to strong stretching force limits (trumpet, stem-flower, and strongly stretched) regimes [7, 8, 9].

Following Kantor and Kardar, the origin of the first term is $\langle \tau \rangle \sim \langle R_g \rangle / v_{CM} = N^{\nu}/N = N^{1+\nu}$ [15], where $\langle R_g \rangle$ is the radius of gyration of the chain. The second term $\eta_{pore} N$ in Eqn. 5 is the contribution of the pore friction η_{pore} and proportional to the contour length of the chain. When the chain length is small, then the pore friction term has a significant effect on the effective translocation exponent α and $1 < \alpha < 1 + \nu$ [12, 13, 14]. This is the reason for the smaller value of the effective translocation exponents 1.44 and 1.50 for the range of $100 < N < 500$ rather than the $\alpha = 1 + \nu = 1.75$ (2D). In the long chain limit $\alpha \to 1 + \nu$, as the dominant contribution to the translocation time comes from the friction due to the movement of the polymer inside the solvent.

Thus, for the case of a semi-flexible chain ($\kappa \neq 0$), a plausible generalization of Eqn. 5 is

$$\langle \tau \rangle \sim [A' \ell_p^{1/(D+2)} N^{1+\nu} + \tilde{\eta}_{pore}(\ell_p) N] \left|\Delta F_{LR}\right|^{-1} \sim N^{\alpha}, \tag{6}$$

where, following Nakanishi [16] and Schaeffer, Pincus and Joanny [17] we have used the generalized expression for the radius of gyration

$$\sqrt{\langle R_g \rangle} \sim \ell_p^{1/(D+2)} N^{\nu} \tag{7}$$

for a semiflexible chain of persistence length ℓ_p in D physical dimensions. Previously we have shown that in 2D Eqn. 7 holds for $L/\ell_p > 1$ [18]. Our combinations of L and ℓ_p ($14 < L/\ell_p < 29$) reported here satisfy this condition well. The second term $\tilde{\eta}_{pore}(\ell_p)$ is now is a function of the chain persistence length. Thus for a given persistence length ℓ_p, Eqn. 6 reduces to Eqn. 5 with $A = \ell_p^{1/d+2} A'$, which then can be used to predict the behavior of longer semiflexible chains. The MFPT for shorter chains is used to numerically obtain A and $\eta_{pore}(\ell_p)$, which then can be

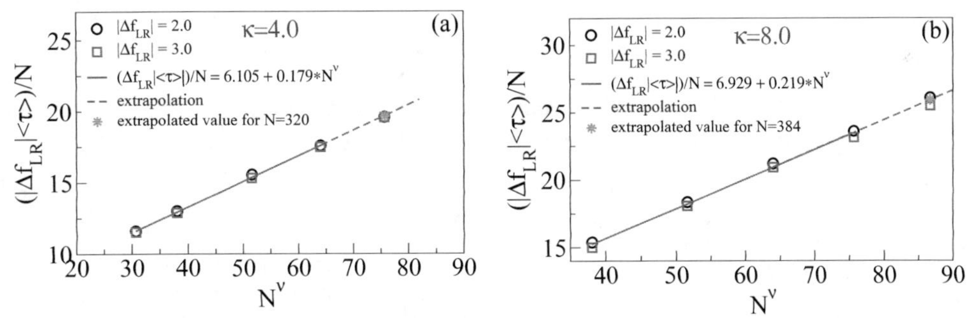

Figure 3. Plots of $\langle \tau \rangle \left|\Delta f_{LR}\right| / N$ as a function of N^{ν} for $\left|\Delta f_{LR}\right| = 2$ (black circles) and 3 (red squares) (a) for $\kappa = 4.0$ and (b) for $\kappa = 8.0$ respectively. In each figure, the green solid line is a straight line fit through the points (which are average values of black circles ($\left|\Delta f_{LR}\right| = 2$) and red squares ($\left|\Delta f_{LR}\right| = 3$) and returns the values of A and η_{pore} respectively. The dotted lines are the extrapolation to larger values of N. The brown stars correspond to the predicted values from this extrapolation for $N = 320$ and $N = 384$ for $\kappa = 4.0$ and 8.0 respectively, while the black circles and red squares are the simulation data.

substituted into Eqn. 5 to predict the MFPT of longer chains. This scheme is shown in Fig. 3. Eqn. 5 can be rewritten as follows:

$$\frac{\langle \tau \rangle \left|\Delta f_{LR}\right|}{N} = A N^{\nu} + \eta_{pore}(\ell_p) \tag{8}$$

Thus for a given value of the chain persistence length ℓ_p, a plot of $\langle\tau\rangle\,|\Delta f_{LR}|\,/N$ as a function of N^ν will produce a straight line with the slope A and η_{pore} as the intercept. Fig. 3 clearly shows that Eqn. 5 is valid for the DNP system, as a linear regression perfectly fits the simulation data and that the data for different values of the bias $|\Delta f_{LR}| = 2$ and 3 collapse on the same master plot. Once η_{pore} is known, one can also check that subtracting the pore friction contribution from $\langle\tau\rangle$ provides a slope of $1+\nu$. We have checked (not shown here) that a log-log plot of $|\Delta f_{LR}| - \eta_{pore}N$ versus N indeed produces a slope $\approx 1+\nu$. Furthermore, that the value of η_{pore} increases with the increasing chain stiffness is consistent with the fact that a stiffer chain translocates slower [19]. This proves that in the limit $d_{LR}/L << 1$ the scaling ansatz of Eqn. 6 works for the DNP system and can be used to predict the translocation time for longer chains.

4. Summary and Conclusion

We have generalized the scaling ansatz for the driven SNP translocation for DNP system and validated the ansatz by carrying out BD dynamics simulation. The ansatz can be used to obtain translocation time for much longer chains for better comparison with experimental systems. Our results will be useful to construct a theoretical platform for the DNP translocation.

5. Acknowledgments

I acknowledge use of computing resources under the auspices of UCF's high performance computing cluster STOKES where all the computations were done, and sincerely thank Prof. Walter Reisner for discussions on various aspects of DNP translocation.

References

[1] For recent reviews in the field please see M. Muthukumar *Polymer Translocation* (CRC Press, Boca Raton, 2011); A. Milchev, J. Phys. Condens. Matter **23**, 103101 (2011); D. Panja, J. Stat. Mech. P06011 (2010); Vladimir V. Palyulin, Tapio Ala-Nissila, and Ralf Metzler, Soft Matter **10**, 9016 (2014); M. Venkatesan and R. Bashir, Nature Nanotechnology **6**, 615–624 (2011).

[2] J. J. Kasianowicz, E. Brandin, D. Branton and D. W. Deamer, *Proc. Natl. Acad. Sci. U.S.A.* **93**, 13770 (1996); A. Meller, L. Nivon, E. Brandin, J. A. Golovchenko, and D. Branton, *ibid* **97**, 1079 (2000). A. Meller, L. Nivon, and D. Branton, *Phys. Rev. Lett.* **86**, 3435 (2001); A. Meller and D. Branton, *Electrophoresis* **23**, 2583 (2002).

[3] S. Pud, S. Chao, M. Belkin, D. Verschureren, T. Huijben, C. van Engelenburg, C. Dekker, and A. Aksimentiev, Nano Lett **16**, 8021-8028 (2016).

[4] Yuning Zhang, Xu Liu, Yanan Zhao, Jen-Kan Yu, Walter Reisner, and William B. Dunbar, Small **14**, 1801890 (2018); Xu Liu, Yuning Zhang, Roland Nagel, Walter Reisner, William B. Dunbar, *ibid*, **15**, 1901704 (2019).

[5] Xu Liu, Philip Zimny, Yuning Zhang, Ankit Rana, Roland Nagel, Walter Reisner, and William B. Dunbar, Small **16**, 1905379 (2019).

[6] Paolo Cadinu *et al.* Nano Lett. **17**, 6376 (2017); *ibid* **18**, 2738 (2018).

[7] T. Ikonen T, A. Bhattacharya, T. Ala-Nissila T, and W. Sung *J. Chem. Phys.* **137**, 085101 (2012).

[8] T. Ikonen, A. Bhattacharya, T. Ala-Nissila and W. Sung, Phys. Rev. E **85**, 051803 (2012).

[9] T. Ikonen, A. Bhattacharya, T. Ala-Nissila and W. Sung, *EPL* **103** 38001 (2013).

[10] G. S. Grest and K. Kremer, Phys. Rev. A **33**, 3628(R) (1986).

[11] S. Bernier S, A. Huang, W. Reisner, and A. Bhattacharya *Macromolecules* **51**, 4012 (2018).

[12] A. Bhattacharya, Computer Simulation Studies in Condensed Matter Physics XXII, Eds. D. P. Landau, S. P. Lewis, and H. B. Schuttler, Elsevier, Physics Proceedia **3**, 1411 (2010).

[13] K. Luo, S. Ollila, I. Huopaniemi, T. Ala-Nissila, P. Pomorski, M. Karttunen, S-C. Ying, and A. Bhattacharya, Phys. Rev. E. **78**, 050901(R) (2008).

[14] A. Bhattacharya, W.H. Morrison, K. Luo, T. Ala-Nissila, S.-C. Ying, A. Milchev and K. Binder, Eur. Phys. J. E **29**, 423-429 (2009).

[15] Y. Kantor and M. Kardar, *Phys. Rev. E* **69**, 021806 (2004).

[16] H. Nakanishi, J. Phys. **48**, 979 (1987); J. Moon and H. Nakanishi, Phys. Rev. A **44**, 6427 (1991).

[17] D. W. Schaefer, J. F. Joanny, and P. Pincus, Macromolecules **13**, 1280 (1980).

[18] A. Huang, A. Bhattacharya, and K. Binder, *J. Chem. Phys.* **140**, 214902 (2014).

[19] Adhikari R. and Bhattacharya A. *J. Chem. Phys.* **138**, 204909 (2013).

On the comparison of optimization algorithms for the random-field Potts model

Manoj Kumar and Martin Weigel
Centre for Fluid and Complex Systems, Coventry University, CV1 5FB, England
E-mail:
manojkmr8788@gmail.com
Martin.Weigel@coventry.ac.uk

Abstract. For many systems with quenched disorder the study of ground states can crucially contribute to a thorough understanding of the physics at play, be it for the critical behavior if that is governed by a zero-temperature fixed point or for uncovering properties of the ordered phase. While ground states can in principle be computed using general-purpose optimization algorithms such as simulated annealing or genetic algorithms, it is often much more efficient to use exact or approximate techniques specifically tailored to the problem at hand. For certain systems with discrete degrees of freedom such as the random-field Ising model, there are polynomial-time methods to compute exact ground states. But even as the number of states increases beyond two as in the random-field Potts model, the problem becomes *NP* hard and one cannot hope to find exact ground states for relevant system sizes. Here, we compare a number of approximate techniques for this problem and evaluate their performance.

1. Introduction

The presence of quenched impurities in magnetic systems can lead to fundamental shifts in material properties including topological changes in the phase diagrams such as the absence of long-range order, especially if frustration is at play [1]. The nature of the quenched average as well as the inherently complex free-energy landscape with a multitude of metastable states separated by barriers observed for systems with strong disorder such as random-field problems and spin glasses quickly push conventional simulation techniques to and beyond their limits [2]. Generalized-ensemble methods including, for instance, multicanonical simulations [3], parallel tempering [4] and population annealing [5–7], do improve the situation substantially in this respect, but the run times required to equilibrate samples continue to rise rapidly with system size.

The same observation applies, in general, to the task of finding ground states of such systems, which is an *NP* hard problem for many discrete spin models and has similar, exponential time complexity for systems with continuous degrees of freedom. Specialist techniques borrowed from theoretical computer science can help alleviate such problems, however [8]. For the Ising spin glass, samples defined on planar graphs can be solved in a time increasing as a polynomial of the system size [9], while the problem on non-planar and, in particular, higher-dimensional lattices is *NP* hard [10]. The ground-state calculation for the random-field Ising model (RFIM), on the other hand, is equivalent to a minimum-cut problem which (in the absence of negative capacities that would occur for the spin glass) is in turn equivalent to the task of finding the maximum

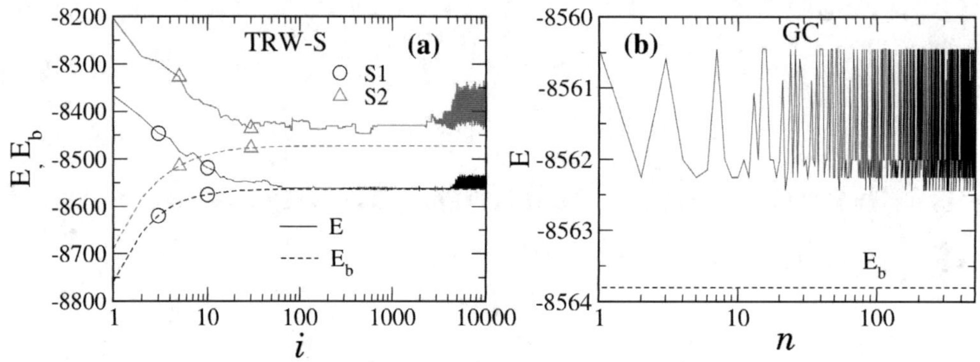

Figure 1. (a) Estimated ground-state energy E and lower bound E_b of two samples ($S1$ and $S2$) of the RFPM from TRW-S as a function of iteration number i. (b) Lowest energy E of $S1$ as estimated from GC calculations using n initial conditions. The dashed line corresponds to the best lower bound for $i = 10\,000$ in (a). Both disorder samples are for a 64×64 lattice with $q = 3$ and $\Delta = 1$.

flow through an auxiliary network connecting two terminal nodes [8, 11]. Powerful methods exist for such problems [12–14] that allow for exact calculations for large systems in any space dimension, reaching up to about 10^7 spins in recent applications [15, 16].

These cases are benign exceptions, however, and most model extensions beyond such solvable cases quickly lead back to exponentially hard tasks. For the random-field problem, for instance, the extension from the twofold Ising to a q-fold Potts symmetry results in a multi-terminal flow problem that can be shown to be *NP* hard [17]. To handle such systems, one might relax the demand of finding exact ground states and revert to stochastic approaches that provide approximate global minima with moderate effort. We have recently shown that this is possible for the case of the *random-field Potts model* (RFPM) with recourse to suitable generalizations of the *graph-cut* (GC) methods used for the RFIM [18]. This approach relies on techniques initially developed for problems in computer vision [17]. While this works quite well, there are some more recent developments in combinatorial optimization that deserve consideration as well, such as the *sequential tree-reweighted message passing* approach of Refs. [19, 20]. In the present note we compare the performance of these methods for finding ground states of the RFPM.

2. Model and methodology

We focus on the RFPM as defined by the following Hamiltonian [21],

$$\mathcal{H} = -J \sum_{\langle ij \rangle} \delta_{s_i,s_j} - \sum_i \sum_{\alpha=0}^{q-1} h_i^\alpha \delta_{s_i,\alpha}, \tag{1}$$

where $\delta_{x,y}$ is the Kronecker delta function. Here, the spins $s_i \in \{0, 1, \ldots, q-1\}$ and $\{h_i^\alpha\}$ is the quenched random field at site i and for component α that is drawn from a normal distribution of mean zero and standard deviation Δ, i.e., $h_i^\alpha \sim \mathcal{N}(0, \Delta)$, such that Δ defines the strength of disorder. The problem of finding a ground state for this Hamiltonian is *NP* hard in general [17] such that in practise one needs to revert to approximation techniques. Such methods have been considered quite extensively in computer vision, where optimization problems very similar to the RFPM ground-state problem occur in the context of image restoration, motion and stereo [14, 17, 19].

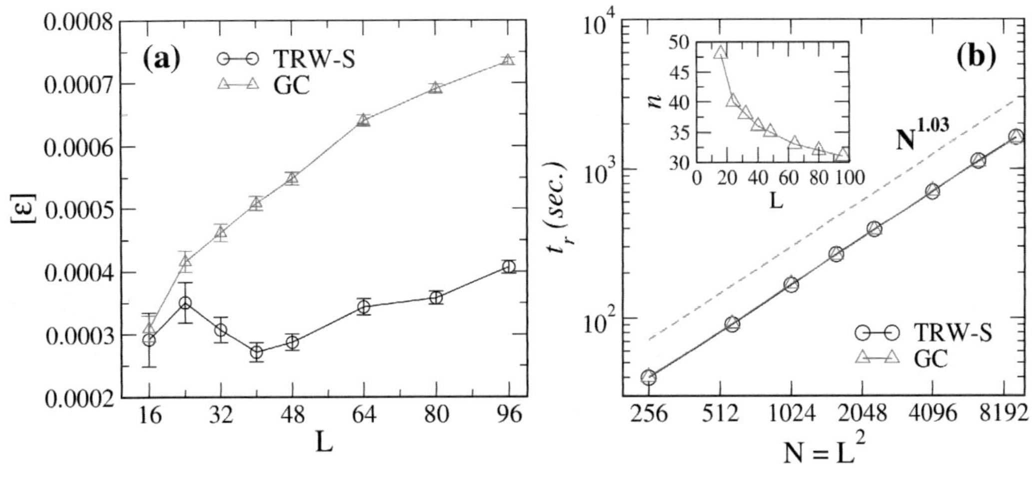

Figure 2. (a) Accuracy $[\varepsilon]$ according to Eq. (2) averaged over 1000 disorder samples against system size L after fixing $i = 1000$ for TRW-S and varying n of GC to match the run time t_r of TRW-S plotted in panel (b). The dashed line corresponds to the power-law increase $\sim N^{1.03}$ resulting from a fit to the data. The inset shows the number of initial conditions used for GC to lead to run time t_r.

One of the most popular approaches was proposed by Boykov *et al.* [17] based on the application of the graph-cut method to embedded binary decision problems. They considered two variant algorithms dubbed α-β-swap and α-expansion, respectively, that are applied to images composed of pixels carrying one of q labels (such as colors, for example). For the α-β-swap, two labels $\alpha \neq \beta \in \{0, 1, ..., q-1\}$ are picked and all other labels are frozen; the update consists of a swap of the labels between regions. In contrast, for α-expansion one picks a label α and attempts to expand it while freezing all the remaining ones, cycling through the labels in turn in q iterations. In both cases, the resulting embedded binary (Ising type) problems are solved exactly using the established techniques for maximum-flow [12–14]. For our application to the RFPM we focus on the α-expansion move which we have shown previously to be quite efficient for finding approximate ground states of this system [18]. By construction, this approach is not guaranteed to result in a ground state, but it will normally lead to a metastable configuration. The approach is stochastic in that the final state depends on the chosen initial spin configuration. As a consequence, results can be systematically improved by performing n independent runs with random initial conditions and picking the result of lowest energy [18, 22]. Since α-expansion cannot increase the energy, this approach is guaranteed to result in the exact answer if (but not only if) the ground state was among the initial configurations. While this appears unlikely for non-trivial system sizes, it indicates that such a procedure should converge in the limit $n \to \infty$.

A different class of approximate methods was considered by Kolmogorov in Ref. [19]. It is based on ideas relating to belief propagation and the more general concept of message passing [23], where "messages" with weights updated according to the local marginal distributions are passed along edges of a graph to iterate probabilistic graphical models until convergence is achieved. Such techniques are exact for trees, but do not guarantee global minima in the presence of loops. The method of (sequential) *tree-reweighted message passing* (TRW-S) proposed in Ref. [19] uses a specific way of covering the graph with trees to approximate the solution. As a byproduct, based on a relaxation for a dual linear programming problem, it also maintains a

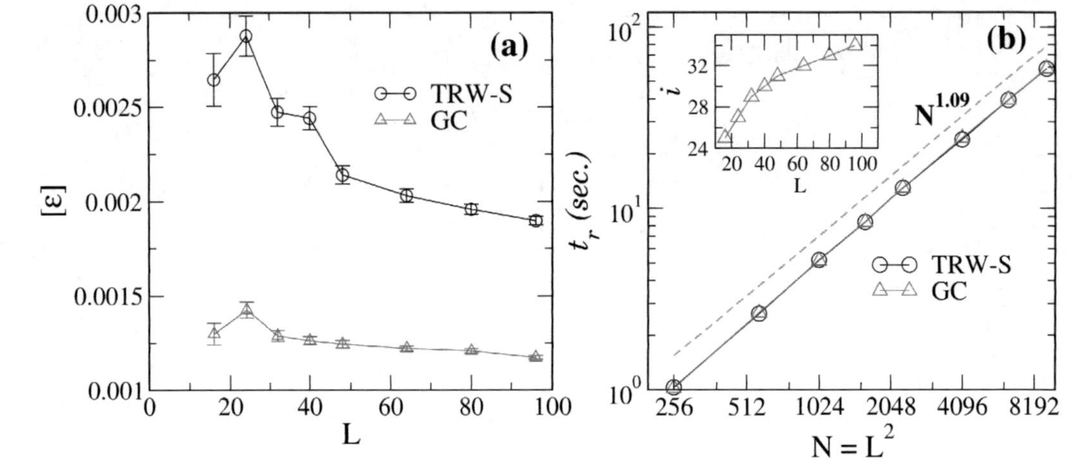

Figure 3. (a) Disorder averaged accuracy $[\varepsilon]$ as a function of L while fixing $n = 1$ of GC and varying i of TRW-S. (b) Run times for this setup. The inset shows the number of TRW-S iterations required. The green dashed line corresponds to the run-time scaling of GC, $t_r \sim N^{1.09}$.

(rigorous) lower bound E_b to the value of the cost function. TRW-S improves with the number of iterations employed, but in contrast to GC it is not known how to systematically improve it by using repeated runs [24]. In most cases, the proposed solution has higher energy than the lower bound, but it is clear that if the two energies coincide an exact ground state has been found. In practise, we run TRW-S for a large number of iterations i, and pick the state of lowest energy encountered.

3. Numerical results

We studied the two-dimensional RFPM for $q = 3$ on a periodic square lattice of edge length $16 \leq L \leq 96$ in order to assess and compare the performance of the GC and TRW-S algorithms discussed above. In the following we focus on $\Delta = 1$, corresponding to quite strong disorder for this 2D system [18]. As demonstrated in Ref. [18], such sizes are large enough for a non-trivial benchmark. For a given disorder sample $\{h_i^\alpha\}$, we vary the number of iterations i of TRW-S (up to a maximum of 10 000), and the number n of random initial spin configurations for GC (up to a maximum of 1000). This is illustrated in Fig. 1 for two samples of size $L = 64$. In TRW-S (panel a) while the energy of the current configuration fluctuates, the value of the lower bound increases monotonically, consistent with Ref. [19]. For GC, on the other hand, the value of the lowest energy found in n independent runs fluctuates and decreases only slowly with n (panel b).

To compare the two approaches more systematically, we tuned both algorithms to result in equal run times, which can be achieved by fixing the number of iterations i of TRW-S or varying the number of initial conditions n of GC, respectively. Performance is assessed by considering the accuracy parameter

$$\varepsilon = \frac{E_b - E_{\min}}{E_b}, \tag{2}$$

where $E_{\min}$ is the best estimate of the ground-state energy from TRW-S or GC, and E_b is the largest value of the lower bound to the ground-state energy of a sample found by TRW-S (which is achieved at $i = 10\,000$). In Fig. 2(a) we compare the disorder averaged accuracy $[\varepsilon]$ between GC and TRW-S for a range of system sizes. Here, the GC method is tuned in n to

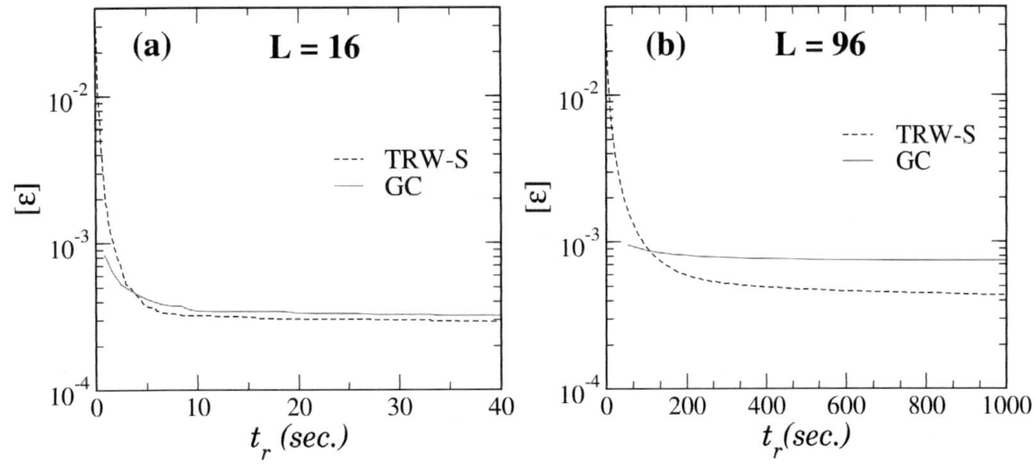

Figure 4. Accuracy $[\varepsilon]$ averaged over 1000 samples as a function of run time t_r for TRW-S and GC. Panel (a) is for $L = 16$ whereas panel (b) is for $L = 96$.

match the run time t_r of TRW-S for $i = 1000$ as shown in panel (b) and its inset. As it is seen there, the total run time is nearly linear in the system volume, a power-law fit resulting in an estimate $t_r \sim N^{1.03}$. The inset shows how the number of initial conditions n required to lead to identical run time with TRW-S depends on L. We also considered a complementary setup, where $n = 1$ was fixed for GC and the number i of iterations of TRW-S was tuned to result in the same run time. The result is shown in Fig. 3. The behavior of run times shown in panel (b) is clearly consistent with a linear scaling in N, in line with previous findings in Ref. [18]. From the comparison of $[\varepsilon]$ shown in panel (a) we see that, interestingly, GC outperforms TRW-S at $n = 1$, meaning that the GC technique finds a quick approximate solution.

Finally, it is instructive to consider the quality of approximation as a direct function of the run time invested. In Fig. 4, we show the comparison of $[\varepsilon]$ for a fixed L but with increasing t_r. It is apparent that initially GC finds better solutions than TRW-S, but with increasing time investment there is a crossover and eventually TRW-S outperforms GC.

4. Summary and conclusions

We studied the performance of the TRW-S and α-expansion GC methods for the two-dimensional random-field Potts model. Both techniques have been tuned to match their run times. Such tuning can be done in two possible ways, i.e., either by varying the number of initial conditions n of GC to match the run time t_r for i iterations of TRW-S, or vice versa. For a fixed number of iterations for TRW-S resp. a fixed number of initial conditions for GC, the run time increases linearly with the number of spins. Investigating both types of comparisons, we find that GC is quicker in finding a reasonable approximation in a short time, but TRW-S leads to better overall results when more significant time is invested. These findings are consistent with the study of Kolmogorov [19], who compared these techniques for a stereo matching problem. While hence TRW-S could be considered the preferable technique, an important advantage of GC is that it converges for $n \to \infty$, albeit slowly [25]. A similar extrapolation parameter is not known for TRW-S [24], suggesting an interesting avenue for future research.

Acknowledgments
The authors acknowledge support by the Royal Society–SERB Newton International Fellowship (NIF\R1\180386). We acknowledge the provision of computing time on the parallel compute cluster Zeus of Coventry University.

References
[1] Young A P (ed) 1997 *Spin Glasses and Random Fields* (Singapore: World Scientific)
[2] Janke W (ed) 2007 *Rugged Free Energy Landscapes — Common Computational Approaches to Spin Glasses, Structural Glasses and Biological Macromolecules* (*Lect. Notes Phys.* vol 736) (Berlin: Springer)
[3] Berg B A and Neuhaus T 1992 *Phys. Rev. Lett.* **68** 9–12
[4] Hukushima K and Nemoto K 1996 *J. Phys. Soc. Jpn.* **65** 1604–1608
[5] Hukushima K and Iba Y 2003 *AIP Conf. Proc.* **690** 200–206
[6] Machta J 2010 *Phys. Rev. E* **82** 026704
[7] Barash L Y, Weigel M, Borovský M, Janke W and Shchur L N 2017 *Comput. Phys. Commun.* **220** 341–350
[8] Hartmann A K and Rieger H 2002 *Optimization Algorithms in Physics* (Berlin: Wiley)
[9] Bieche I, Maynard R, Rammal R and Uhry J P 1980 *J. Phys. A* **13** 2553
[10] Barahona F 1982 *J. Phys. A* **15** 3241
[11] Anglès d'Auriac J C, Preissmann M and Rammal R 1985 *J. Physique Lett.* **46** L173
[12] Ford L R and Fulkerson D R 1962 *Flows in Networks* (Princeton, NJ: Princton University Press)
[13] Goldberg A V and Tarjan R E 1988 *J. ACM* **35** 921–940
[14] Boykov Y and Kolmogorov V 2004 *IEEE T. Pattern Anal.* **26** 1124–1137
[15] Stevenson J D and Weigel M 2011 *Europhys. Lett.* **95** 40001
[16] Fytas N G and Martín-Mayor V 2013 *Phys. Rev. Lett.* **110** 227201
[17] Boykov Y, Veksler O and Zabih R 2001 *IEEE Trans. Pattern Anal. Mach. Intell.* **23** 1222–1239
[18] Kumar M, Kumar R, Weigel M, Banerjee V, Janke W and Puri S 2018 *Phys. Rev. E* **97** 053307
[19] Kolmogorov V 2006 *IEEE Trans. Pattern Anal. Mach. Intell.* **28** 1568–1583
[20] Kolmogorov V 2014 *IEEE Trans. Pattern Anal. Mach. Intell.* **37** 919–930
[21] Blankschtein D, Shapir Y and Aharony A 1984 *Phys. Rev. B* **29** 1263–1267
[22] Weigel M 2007 *Phys. Rev. E* **76** 066706
[23] Mézard M and Montanari A 2009 *Information, Physics, and Computation* (Oxford: Oxford University Press)
[24] Kolmogorov V Private communication
[25] Kumar M and Weigel M In preparation

Visualization by optical fluorescence of two-phase flow in a three-dimensional porous medium

Joachim Falck Brodin[1], Marcel Moura[1], Renaud Toussaint[2], Knut Jørgen Måløy[1] and Per Arne Rikvold[1,3]

[1] PoreLab, NJORD Centre, Department of Physics, University of Oslo, P.O. Box 1048 Blindern, 0316 Oslo, Norway
[2] Université de Strasbourg, CNRS, Institut de Physique du Globe de Strasbourg, UMR 7516, France
[3] Department of Physics, Florida State University, Tallahassee, FL 32306-4350, USA

E-mail: j.f.brodin@fys.uio.no, marcel.moura@fys.uio.no, renaud.toussaint@unistra.fr, k.j.maloy@fys.uio.no, prikvold@fsu.edu

Abstract. Slow flow of a single fluid through a porous medium is well understood on a macroscopic level through Darcy's law, a linear relation between flow rate and a combination of pressure differences, viscosity, and gravitational forces. Two-phase flow is complicated by the interface separating the fluids, but understanding of two-dimensional, two-phase flow has been obtained from experiments using transparent cells. In most three-dimensional media, however, visual observation is difficult. Here, we present preliminary results of experiments on a model medium consisting of randomly packed glass spheres, in which one fluorescent liquid invades another. By refractive index matching and scanning with a sheet-shaped laser beam, we obtain slices of the flow patterns, which we combine into three-dimensional pictures. We observe a compact region of invading fluid, surrounded by finger-like protrusions. The compact region becomes more dominant with increasing invader flow rate. The patterns are theoretically analyzed in terms of the interplay between gravitational, viscous, and capillary forces.

1. Introduction

Many systems, from host rocks in aquifers and oil fields to diapers, kitchen sponges, and biological tissues, can be described as porous materials. The behavior of fluids in porous media has been under quantitative study at least since the 19th century, when the French water-works engineer, Henry Darcy, studied the flow of water through sand-filled pipes [1]. Macroscopically, slow ("creeping") flow of a single fluid through a porous medium is well described by his results. They are summarized by "Darcy's Law," a linear-response relation between fluid flow rate, pressure differences, viscosity, and gravity. Applied to a vertical pipe of constant cross-sectional area A, into which fluid enters with a constant, downward flow rate Q at $z = 0$, it can be written in modern notation as

$$p(z) = p(0) - \left(\rho g - \frac{\mu}{k}\frac{Q}{A}\right) z \, . \tag{1}$$

Here, $p(z)$ is the pressure at z, ρ is the fluid density, g is the acceleration of gravity, μ is the fluid viscosity, and k is known as the *permeability*. The latter is a characteristic of the medium, which can be considered as an effective cross-sectional area of the channels through the microscopic pore structure. The ratio Q/A is known as the *filtration velocity*.

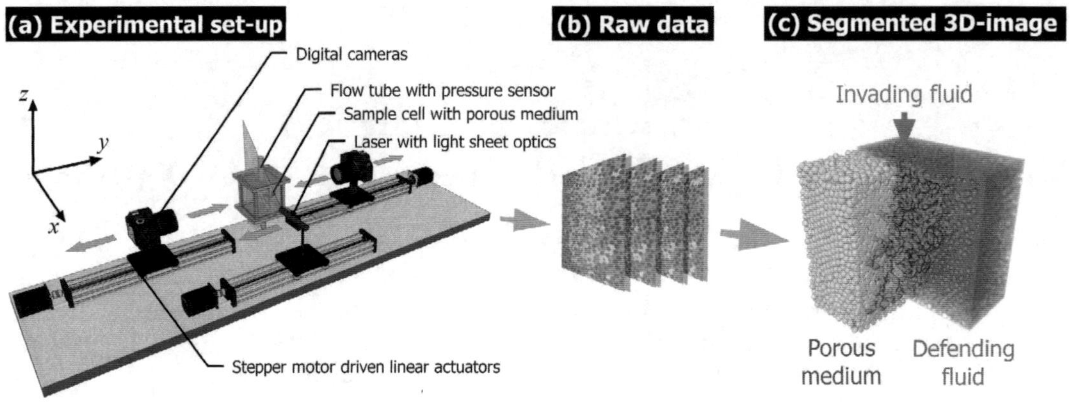

Figure 1. (a): The experimental set-up to produce 3D images of two-phase flow in a porous medium by laser-induced fluorescence. The box between the cameras is the flow cell, containing a random packing of glass spheres. Initially, the medium is saturated with rapeseed oil tagged for red fluorescence, which is later invaded from above by glycerol tagged for green fluorescence. **(b):** Some of the 2D slice images making up the raw data. **(c):** Segmented 3D cut-away image at the moment the glycerol percolates to the bottom. The gray section shows the porous medium, the red section shows the oil, and the green object is the invading glycerol body.

Two-phase flow is complicated by the capillary forces at the interface separating the fluids. Understanding of the patterns arising from competing gravitational, viscous, and capillary forces for two-dimensional (2D) two-phase flow [2, 3, 4, 5] has been obtained using transparent plates confining a thin layer of the porous medium, enabling visual observation of the flow. Such observation is difficult in three dimensions (3D).

In Sec. 2, we present preliminary results from our experiments on a 3D model medium consisting of randomly packed glass spheres, in which one fluorescent fluid is invaded by another. In Sec. 3, the observations are analyzed and semi-quantitatively explained, using an approximate generalization of Eq. (1) that describes the effects of competing forces in the 3D flow, analogous to previous 2D results [2, 3, 4, 5].

2. Experimental set-up and results

We consider a situation, in which a dense, viscous liquid (glycerol, $\rho_1 = 1.255$ g/ml, $\mu_1 = 1.350$ Pa s) is injected from above into a porous medium, which is initially saturated with a less dense and much less viscous liquid (rapeseed oil, $\rho_2 = 0.912$ g/ml, $\mu_2 = 0.058$ Pa s).

The experimental set-up is illustrated in Fig. 1. An $80 \times 80 \times 80$ mm glass box with inlet and outlet pipes (radius $r_{\text{in}} = 1.5$ mm) at top and bottom, respectively, is filled with a random packing of glass spheres (radius $r_0 = 1.5$ mm and measured porosity $\phi = 0.41$). Initially, the pore space between the spheres is saturated with the oil tagged with a red, fluorescent dye. Then, glycerol (tagged for green fluorescence) is injected through the inlet at controlled, constant flow rates between 0.1 and 1.0 ml/min, until the invading glycerol reaches the bottom (percolation).

The 3D flow patterns are visualized by scanning a vertical, sheet-shaped laser beam across the sample at 3.5 mm/s, recording 2D "slice" images of the fluorescence with digital cameras at a rate of approximately 49 frames/s. Optical transparency is ensured by matching the refractive indices of the fluids and the medium [6, 7]. The cameras are moved horizontally, so as to maintain constant optical distance between each camera and the light sheet. 3D images and movies are constructed from the 2D slices, using the Amira-Avizo software from Thermo Scientific.

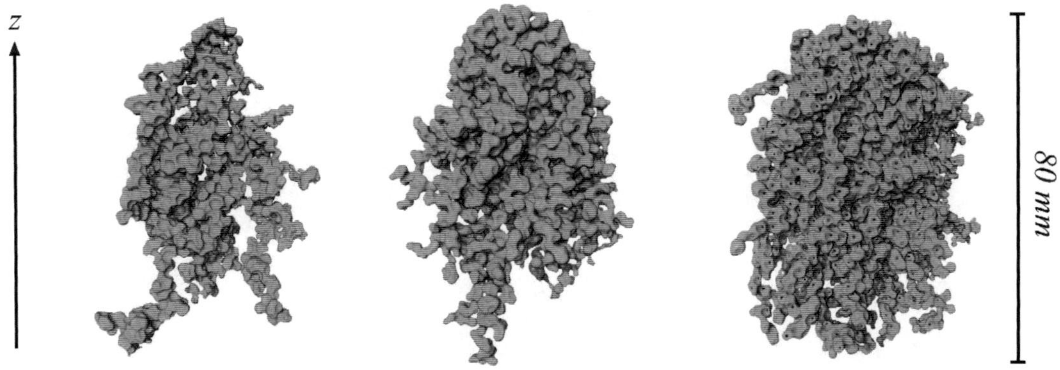

Figure 2. Images of the 3D configuration of the invading glycerol liquid at percolation. Flow rates, from left to right: $Q = 0.1$, 0.5, and 1.0 ml/min. The structures consist of a compact region, surrounded by thin "fingers." In these pictures, the latter are best seen in the lower part and at the sides of the body of glycerol. The compact centers grow with increasing Q, and also with time.

Images of the invading glycerol configuration at percolation, produced at different flow rates Q, are shown in Fig. 2. We observe that the configurations are more compact near the center, while extending long "fingers" near the periphery. The compact region grows as Q is increased.

3. Theoretical analysis

To estimate the distance below the inlet, where the interface becomes unstable with respect to gravitational fingering, we introduce a simple, 3D generalization of Darcy's law, Eq. (1). In this picture, we approximate the filtration velocity of the invading fluid as uniform over a hemispherical surface, centered at the inlet. For simplicity, we ignore boundary conditions at the outlet, walls, and floor of the box.

The pressure difference between the two fluids at their interface is known as the *capillary pressure difference*. Its value across the interface at a point a distance R from the inlet and at a depth $-z > 0$, (R, z), is estimated by integrating the Darcy equation along two paths from the inlet to (R, z), each lying entirely inside one of the fluids.

$$\Delta p(R,z,t) \approx \Delta p(0,0,t) - zg\Delta\rho - \frac{\mu_1 Q}{2\pi k}\int_{r_{\rm in}}^{R}\frac{\mathrm{d}r}{r^2} = \Delta p(0,0,t) - zg\Delta\rho + \frac{\mu_1 Q}{2\pi k}\left[\frac{1}{R} - \frac{1}{r_{\rm in}}\right], \quad (2)$$

where $\Delta\rho = \rho_1 - \rho_2$, and $0 \leq r_0 \leq -z \leq R$. The permeability for this medium is estimated from the Kozeny-Carman relation [7] as $k \approx \frac{r_0^2}{45}\frac{\phi^3}{(1-\phi)^2} = 0.99 \times 10^{-8}\,\mathrm{m}^2$. $\Delta p(0,0,t)$ is the time-dependent pressure applied by the pump at the inlet to maintain the flow rate constant at Q.

For the interface to progress past a point $\vec{R}$ at (R, z), the capillary pressure difference at that point, $\Delta p(R,z,t) = p_1(R,z) - p_2(R,z)$, must exceed a local pressure threshold, $p^{\rm t}(\vec{R})$. Each threshold represents the capillary pressure required for the meniscus (surface tension $\sigma \approx 16.23 \times 10^{-3}$ N/m) to pass from one pore to a neighboring one through a narrow *pore throat* at this location. We assume that these thresholds are independent random variables with a probability density of mean $\langle p^{\rm t}\rangle$ and a width $W^{\rm t}$ of comparable magnitude. Using the Young-Laplace equation, we approximate $W^{\rm t} \approx \langle p^{\rm t}\rangle \approx 2\sigma/r_0 \approx 20\,\mathrm{Pa}$. For simplicity, we use spherical

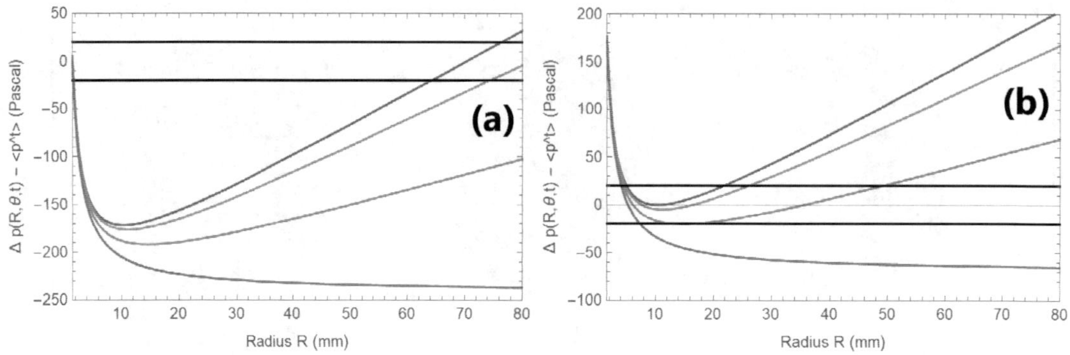

Figure 3. The capillary-pressure function $\Delta p(R,\theta,t) - \langle p^{\mathrm{t}}\rangle$ for $Q = 1.0$ ml/min, vs distance R from the inlet, along straight lines at angles θ with the negative z axis. From top to bottom are $\theta = 0°$ (vertically down), 30°, 60°, and 90° (horizontal). Parts (a) and (b) represent different times, and the horizontal, black lines represent the capillary-pressure interval, $[\langle p^{\mathrm{t}}\rangle - W^{\mathrm{t}}, \langle p^{\mathrm{t}}\rangle + W^{\mathrm{t}}]$. **(a):** At a very early time, when $\Delta p(0,t) - \langle p^{\mathrm{t}}\rangle \approx 0$. The invading glycerol can then only penetrate a very short distance through the largest pores. To drive the flow further at constant Q, the pump increases $\Delta p(0,t)$. **(b):** At a later time, when the flow is just becoming unstable toward gravitational fingering near the saddle point at the minimum for $\theta = 0°$.

coordinates to write $-z = R\cos\theta > 0$, were θ is the angle between a straight line from the inlet and the negative z axis, and rename quantities in Eq. (2) to write the threshold requirement as

$$\Delta p(R,\theta,t) \approx \Delta p(0,0,t) + GR\cos\theta + M\left[\frac{1}{R} - \frac{1}{r_{\mathrm{in}}}\right] > p^{\mathrm{t}}(\vec{R})\,. \tag{3}$$

With experimental values from Sec. 2, $G = 3.36$ Pa/mm and $M = 362$ Pa mm for $Q = 1.0$ ml/min.

In Fig. 3 we plot, at two different times, $\Delta p(R,\theta,t) - \langle p^{\mathrm{t}}\rangle$ vs R for different angles θ. At both times, the two thick, black lines mark a pressure-threshold interval of width $2W^{\mathrm{t}}$, centered at $\langle p^{\mathrm{t}}\rangle$. Part (a) shows an early time, when $\Delta p(0,0,t) \approx \langle p^{\mathrm{t}}\rangle$, and part (b) shows a later time, when the flow is just becoming unstable against gravitational fingering, as discussed below.

The spatially constant terms in Eq. (3) vanish in the gradient of $\Delta p(R,\theta,t)$, which in spherical coordinates is,

$$\nabla[\Delta p(R,\theta,t)] \approx \hat{r}\left(G\cos\theta - \frac{M}{R^2}\right) - \hat{\theta}G\sin\theta\,. \tag{4}$$

The gravitational driving term, $G\cos\theta$, and the viscous drag term, $-M/R^2$, compete in the radial component (proportional to what is known as the "modified Bond number" [3]). The latter dominates near the inlet (negative slopes in Fig. 3), while the former dominates far below the inlet (positive slopes). The gradient vanishes at a point on the negative z axis, with

$$R_0 = -z = \sqrt{M/G}\,. \tag{5}$$

This is a saddle point for the capillary pressure difference. For the compact cluster of invading fluid, R_0 is a critical size, beyond which its interface against the defending fluid becomes unstable toward gravitationally driven finger growth. From Eq. (5), our estimates for R_0 are approximately 10.4 mm for $Q = 1.0$ ml/min, 7.3 mm for 0.5 ml/min, and 3.3 mm for 0.1 ml/min.

In Fig. 4, we show an experimental image and a theoretical capillary-pressure landscape, both projected onto the vertical $(x, 0, z)$ plane. They represent a time near that, at which the

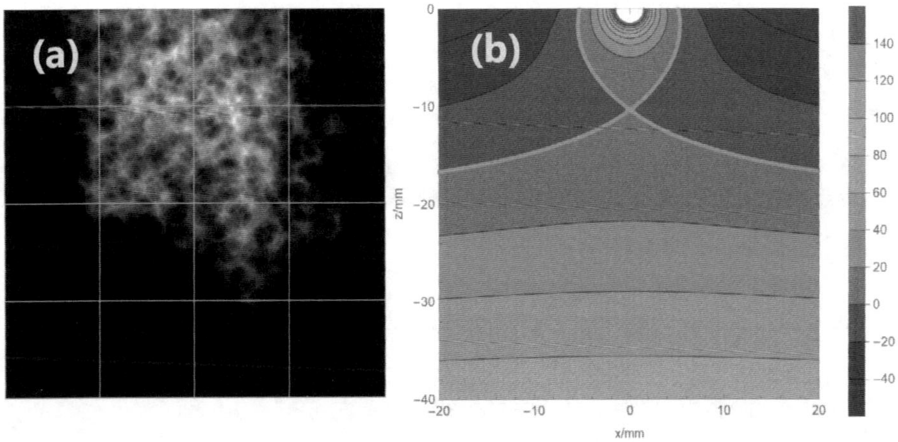

Figure 4. The compact center of invading glycerol for $Q = 1.0$ ml/min near the time it becomes unstable. Inlet at top center of each part. **(a):** Vertical density projection of the experimental result. **(b):** Contour map of the theoretical capillary-pressure landscape in the vertical plane $(x, 0, z)$. Yellow, crossing curves mark the isobar at the saddle point. Other isobars are shown at 20 Pa $\approx W^{\text{t}}$ intervals – blue: negative, green-red: positive. The vertical color bar shows the pressure differences from the saddle point in Pascal. The white semicircle at the top shows the position and diameter of the inlet. Both plots are shown on the same scale of 40 mm $\times$ 40 mm.

flow becomes unstable, as in Fig. 3(b). (This is an earlier time than that depicted in Fig. 2.) Both images illustrate $Q = 1.0$ ml/min and cover the same area of 40 mm $\times$ 40 mm. Part (a) is a density projection of the compact center of the invading glycerol cluster. (Fingers were removed by an Erosion/Dilation method [8].) Part (b) is a contour map of the capillary-pressure landscape. The isobar corresponding to the saddle point is marked by a thick, yellow curve. Other isobars at intervals of 20 Pa $\approx W^{\text{t}}$ are marked by color changes. The growth is expected to be stable in the positive-pressure region surrounded by the closed part of the saddle-point isobar. Comparing Figs. 4(a) and (b), the compact glycerol body is seen to have a diameter of about 20 mm and to extend no more than about ± 20 Pa outside the theoretically expected stable region in the capillary-pressure landscape. Thus, the agreement between the experimental result and our simple theoretical estimate appears satisfactory.

Further refinements of the experiments and theoretical analysis are in progress.

Acknowledgments

We thank G. M. Buendía and A. J. Gurfinkel for useful comments. Supported by the Research Council of Norway through the Center of Excellence funding scheme, Project No. 262644.

References

[1] Darcy H 1856 *Les fontaines publiques de la ville de Dijon* (Paris: Dalmont)
[2] Birovljev A, Wagner G, Meakin P, Feder J and Jøssang T 1995 *Phys. Rev. E* **51** 5911
[3] Méheust Y, Løvoll G, Måløy KJ and Schmittbuhl J 2002 *Phys. Rev. E* **66** 051603
[4] Toussaint R, Måløy KJ, Méheust Y, Løvoll G, Jankov M, Schäfer G and Schmittbuhl J 2012 *Vadose Zone J.* **11** vzj2011-0123
[5] Vasseur G, Luo X, Yan G, Loggia D, Toussaint R and Schmittbuhl J 2013 *Marine Petro. Geol.* **45** 150
[6] Stöhr M, Roth K and Jähne B 2003 *Exp. Fluids* **35** 159
[7] Dalbe M-J and Juanes R 2018 *Phys. Rev. Appl.* **9** 024028
[8] Haralick RM, Sternberg SR and Zhuang X 1987 *IEEE Trans. Pattern Anal. Machine Intel.* **PAMI-9** 532

Fast simulation of a large polymer with untruncated interaction near the collapse transition

Stefan Schnabel and Wolfhard Janke

Institut für Theoretische Physik, Universität Leipzig, IPF 231101, 04081 Leipzig, Germany

E-mail: schnabel@itp.uni-leipzig.de

Abstract. We show how polymer models with untruncated intra-molecular interactions and continuous degrees of freedom can be investigated by means of Monte Carlo techniques even for large systems. To this end we adapt Clisby's method for the simulation of self-avoiding walks and introduce a novel way to implement the well-known Metropolis algorithm.

1. Introduction

In the phase diagram of flexible polymer chains the Θ–point lies in between good and bad solvent. For the good solvent, the system maximizes the polymer–solvent interface area causing the polymer to occupy extended configurations that resemble random walks which are self-avoiding due to the excluded volume interaction. The average end–to–end distance of a chain of length L, therefore, scales like $\langle r_{\mathrm{ee}} \rangle \propto L^{\nu}$ with $\nu \approx 0.588$ in three dimensions. In a bad solvent the surface area is minimized and the polymer forms denser globule-like shapes. The system is expected to grow like a droplet of constant density such that the linear extension is proportional to the third root of the droplet's volume which in turn is proportional to L. Hence, $\langle r_{\mathrm{ee}} \rangle \propto L^{\nu}$ with $\nu = 1/3$. In between at the Θ-point the polymer is supposed to behave like an ideal chain or a Gaussian random walk (GRW) which implies $\nu = 1/2$.

There have been many numerical studies of the behavior of polymers near the Θ–point [1, 2, 3, 4]. In the majority of them lattice polymers were considered since generally such models allow for the investigation of longer chains. For example in a recent publication [5] off-lattice polymers of length $L < 500$ were studied while Vogel et al. [6] simulated lattice polymers with $N = L + 1 = 32000$ monomers. Due to the underlying lattice geometry the latter are less realistic than the former and it is, therefore, desirable to have efficient algorithms for off-lattice polymers at hand. The naïve method to simulate these models is hampered by the large number of terms in the Hamiltonian and requires $\propto L^2$ operations for an individual move. This can relative easily be improved to $\propto L$, e.g. by partitioning of space, but only if the interaction potential is of limited range. Since a number of natural choices like the standard 12-6 Lennard-Jones potential are thus excluded from consideration, a method that is able to also investigate potentials that are non-zero at all distances would be useful.

The paper is organized as follows: In section 2 we briefly discuss the Lennard-Jones polymer model that we are using. Then, in section 3, we describe a parsimonious version of the Metropolis algorithm which allows to decide upon suggested configuration changes without exact knowledge of the change in energy. This is followed by a discussion of the employed data structure, i.e., our version of Clisby's binary tree, in section 4. Some data regarding the performance of our method is presented in section 5, before we show the results of a numerical experiment estimating the Θ-temperature in section 6. Some concluding remarks finish the paper in section 7.

2. Model

We consider a bead–stick polymer model with continuous degrees of freedom in three dimensions. Monomers (beads) are placed at coordinates $\mathbf{x}_1, \ldots, \mathbf{x}_N$ and are connected by bonds of fixed length,

$$\mathbf{b}_i = \mathbf{x}_{i+1} - \mathbf{x}_i, \qquad |\mathbf{b}_i| = b. \tag{1}$$

All monomers interact via a 12–6 Lennard-Jones potential

$$U(r) = 4\epsilon\left[\left(\frac{\sigma}{r}\right)^{12} - \left(\frac{\sigma}{r}\right)^{6}\right], \qquad \sigma = \frac{b}{2^{1/6}} \tag{2}$$

with their diameter σ chosen such that the minimum distance of the interaction equals the bond length b. In the following we will set $\epsilon = 1$ to fix the energy and temperature scales while assuming that $b = 1$ such that all distances are given in units of b. The complete Hamiltonian reads:

$$\mathcal{H} = \sum_{i=1}^{N-1} \sum_{j=i+1}^{N} U(|\mathbf{x}_i - \mathbf{x}_j|) + \sum_{i=1}^{N-1} \delta(|\mathbf{x}_{i+1} - \mathbf{x}_i| - b). \tag{3}$$

In this notation the total number of beads is denoted by N and the length of the polymer is $L = N - 1$. Interactions of adjacent beads are, of course, constant $U(|\mathbf{x}_i - \mathbf{x}_{i+1}|) = -\epsilon$ due to their constant distance and could be omitted. We include them in the Hamiltonian because it leads to a less complicated program code.

3. A parsimonious Metropolis algorithm

One of the first and most widely applied Monte Carlo methods is the Metropolis algorithm [7]. It generates a Markovian chain of microstates μ_i that follow a Boltzmann distribution

$$P(\mu) \propto e^{-\beta E(\mu)}, \tag{4}$$

where $E(\mu) = \mathcal{H}(\mu)/\epsilon$ is the energy associated with the state μ in units of ϵ and $\beta = \epsilon/k_\mathrm{B}T$ is the inverse temperature in natural units. The transition to a new state μ_{i+1} is done in the following way:

- A new state ν is proposed by a random modification to μ_i. Typically, the probability for this proposal is symmetric $P_\mathrm{prop}(\mu \to \nu) = P_\mathrm{prop}(\nu \to \mu)$.
- The energy difference of the two states $\Delta E = E(\nu) - E(\mu_i)$ is determined and the acceptance probability $P_\mathrm{acc} = \min\left(1, e^{-\beta\Delta E}\right)$ is calculated.
- A (pseudo) random number $\xi \in [0, 1)$ is drawn from a uniform distribution.
- The modification is accepted ($\mu_{i+1} := \nu$) if $\xi < P_\mathrm{acc}$ and rejected ($\mu_{i+1} := \mu_i$) otherwise.

This sequence can be shortened if $\Delta E < 0$ (and $\beta \geq 0$) since such an update is always accepted.

If this procedure is followed, ΔE will always be calculated with the maximal accuracy that is possible with the respective data type, e.g., a relative accuracy of about 10^{-15} for 64 bit floating–point numbers. However, when doing the comparison in the last step this accuracy is hardly ever required. This raises the question if it is feasible to allow for less accurate representations of ΔE within the program and whether this can be exploited to accelerate the simulation.

We reformulate the algorithm and start with drawing the random number ξ. The condition

$$\xi < P_{\mathrm{acc}} = \min(1, e^{-\beta\Delta E}) \tag{5}$$

is equivalent to

$$\Delta E < -\frac{\ln\xi}{\beta} \tag{6}$$

again assuming $\beta > 0$. If one is able to establish bounds for the change in energy $\Delta E \in [\Delta E_{\mathrm{min}}, \Delta E_{\mathrm{max}}]$ the decision can be made if either $\Delta E_{\mathrm{max}} < -\frac{\ln\xi}{\beta}$ (accept) or $\Delta E_{\mathrm{min}} \geq -\frac{\ln\xi}{\beta}$ (reject). If neither condition holds the estimate has to improved, i.e., the lower and upper bounds have to be narrowed. This approach will be useful if low-accuracy estimates of the change in energy can be obtained substantially faster than estimates with high accuracy or the exact value itself.

4. Binary Trees

Recently, Clisby [8] has introduced a method employing binary trees to the simulation of self-avoiding walks on periodic lattices and has achieved remarkable success. It is now possible to investigate walks with tens of millions of steps. We have adopted this method for walks in continuous space [9] and use the binary trees here as the fundamental data structure for the representation of the polymer.

The tree's leaves, i.e., nodes at the bottom that are connected to only a single other node, correspond to individual monomers and store this particular monomer's position. Nodes at higher levels represent larger parts of the monomers. If the tree is balanced each node in the second level represents the two monomers that correspond to the children-nodes, nodes in the third level correspond to groups of four monomers again comprised of the monomers from the children-nodes and so on. Our intent is to use these higher-level nodes in order to estimate the interactions between groups of polymers and for that purpose we must select which information are to be stored with them. For our current implementation we opted for a simple selection. The node stores the number of monomers it represents as well as the geometric information of a sphere – center $\mathbf{y}$ and radius r – which contains all those monomers. Additional data like the center of mass or higher moments could be used as well. Note that these data can relatively easily, i.e., with complexity $\mathcal{O}(1)$, be derived from the respective data stored in the children-nodes and that the data of the entire tree can thus be created bottom-up within $\propto N$ operations. An important property of this data structure is that transformations that are to be enacted during a Monte Carlo step can be applied to the higher-level nodes. If the monomers represented by a node are to be rotated collectively, the new position of its sphere can be derived directly from the old one without explicit calculation of the individual new monomer positions.

In order to estimate the interaction between groups of monomers represented by two spheres we exploit that the Lennard-Jones potential (2) is monotonic for $r > b$. If two nodes A and B represent groups of n_A and n_B monomers located in spheres around the points $\mathbf{y}_A$ and $\mathbf{y}_B$ with radii r_A and r_B it is clear that for any monomer $k \in A$ and $l \in B$ it is

$$|\mathbf{y}_A - \mathbf{y}_B| - r_A - r_B < |\mathbf{x}_k - \mathbf{x}_l| < |\mathbf{y}_A - \mathbf{y}_B| + r_A + r_B. \tag{7}$$

For the interaction between the two nodes

$$E_{AB} = \sum_{k\in A}\sum_{l\in B} U(|\mathbf{x}_k - \mathbf{x}_l|) \tag{8}$$

it follows that if $|\mathbf{y}_A - \mathbf{y}_B| - r_A - r_B \geq b$, i.e., if $U(r)$ is monotonic in the relevant interval, we can obtain boundaries $E_{AB} \in [E_{AB}^{\mathrm{min}}, E_{AB}^{\mathrm{max}}]$ given by

$$E_{AB}^{\mathrm{min}} = n_A n_B U(|\mathbf{y}_A - \mathbf{y}_B| - r_A - r_B) \tag{9}$$

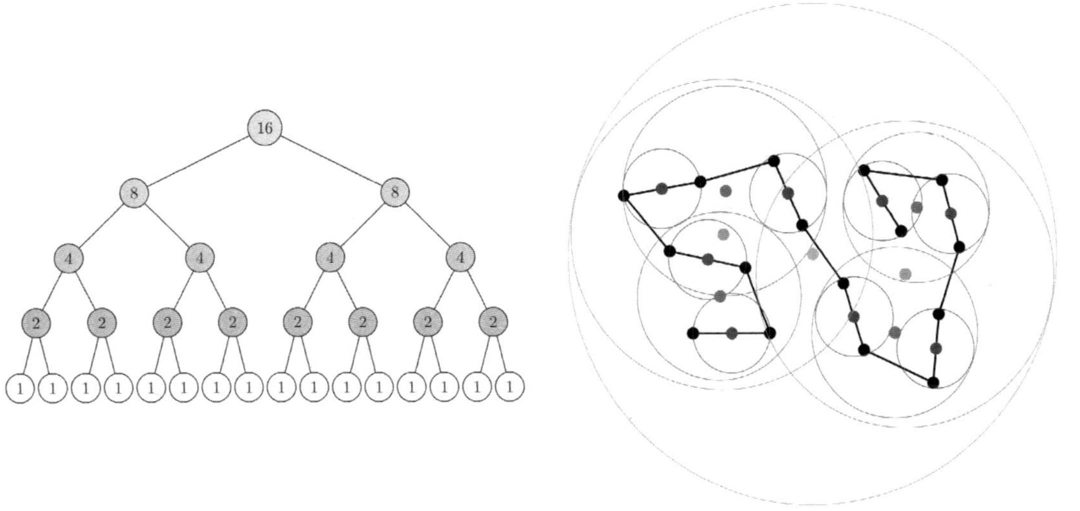

Figure 1. (Left) Binary tree with the numbers of monomers each node represents. (Right) Two-dimensional sketch of the way geometric information from groups of monomers is aggregated as spheres on different levels in the binary tree.

and

$$E_{AB}^{\max} = n_A n_B U(|\mathbf{y}_A - \mathbf{y}_B| + r_A + r_B). \tag{10}$$

If the obtained estimate is not precise enough it can be improved by involving nodes at the next lower level. If the children of node A are labeled A_l and A_r an almost always better[1] estimate is given by

$$E_{AB} = E_{A_lB} + E_{A_rB} \in [E_{A_lB}^{\min} + E_{A_rB}^{\min}, E_{A_lB}^{\max} + E_{A_rB}^{\max}]. \tag{11}$$

In Fig. 1 the binary tree is displayed in its balanced form, where each inner node is parent to two subtrees of equal size. When a Monte Carlo step is performed it can be useful to alter this shape. In our simulation we used two different types of steps. One is the pivot move, where from one fulcrum monomer $\mathbf{x}_p$ on all monomers $\mathbf{x}_k$ with $k \in \{p+1, \ldots, N\}$ are rotated by a random angle around an axis through $\mathbf{x}_p$. The other move is best described using the bond vectors $\mathbf{b}_i$. One of them is randomly changed $\mathbf{b}_{p+1} \to \mathbf{b}'_{p+1}$ while all others remain the same. This amounts to a constant shift $\mathbf{x}_k \to \mathbf{x}_k + \mathbf{b}'_{p+1} - \mathbf{b}_{p+1}$ for all monomers with $k \in \{p+1, \ldots, N\}$. In both cases the polymer decomposes into two parts one of which remains unaltered while the other is modified. It is convenient to rearrange the tree such that each part is represented by an individual node (Fig. 2). This can be done by a sequence of so-called tree–rotations (Fig. 3). Once the tree is in the new shape the change in energy can be estimated by simultaneously estimating the interactions between these two nodes in their pre–update state as well as with the right node's sphere which contains the modified monomers in the post–update position. Both guesses can be improved in parallel until ΔE is known with sufficiently high precision and a decision can be made.

5. Performance

In order to evaluate the performance of the method we performed simulations using our algorithm and a naïve method for different sizes at $T = 4$ which is slightly above T_Θ (see below). The

[1] In rare cases rebalancing in a node's subtree can cause the sphere in one of that node's children to partly lie outside of the node's own sphere even though the monomers are still contained inside it. This can cause bounds of ΔE to become wider instead of narrower.

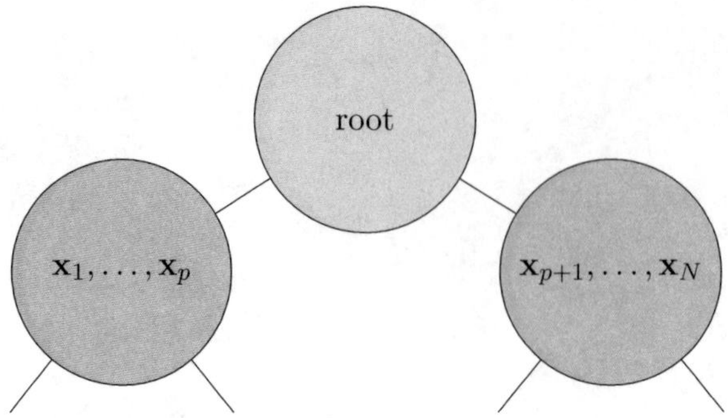

Figure 2. Top part of a binary tree that has been arranged to facilitate an update of the monomers $\mathbf{x}_1, \ldots, \mathbf{x}_p$. Note that as before the displayed nodes do not store the $\mathbf{x}_i$ but the data for a sphere containing them.

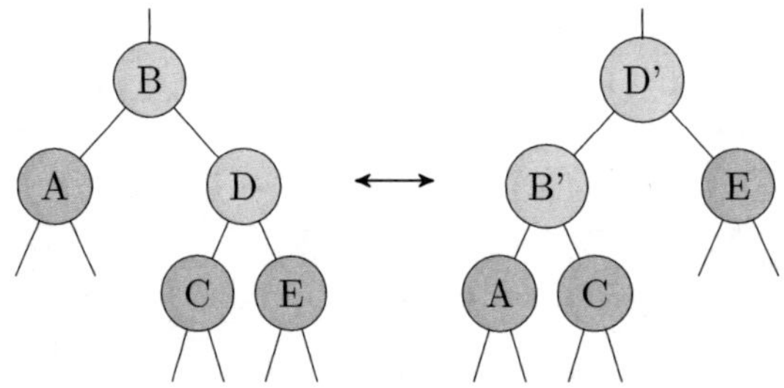

Figure 3. Tree-rotations are used to transform the binary before and after a Monte Carlo step. Only the nodes B and D are being modified and the order from left to right is preserved.

results are shown in Fig. 4. As expected the time required by the standard algorithm grows proportionally with L^2 as soon as the small overhead becomes negligible. In contrast, the more complicated data structures used with our method lead to a slower simulation for very small systems, but it soon ($L > 100$) becomes more efficient when the polymer length is increased. Moreover, the data clearly shows that the time needed for individual updates grows slower than linearly and will for larger sizes than the ones we investigated likely beat $\mathcal{O}(\sqrt{L})$.

6. Critical temperature

At the Θ–point a large enough polymer at large enough scales is supposed to take the shape of a GRW. For smaller chains there might not be a temperature where this behavior is completely realized, but we still use a characteristic of GRW to identify the transition temperature. For a GRW the end–to–end vector $\mathbf{r}_{ee} = \mathbf{x}_{L+1} - \mathbf{x}_1$ is distributed according to a Gaussian distribution

$$P_{3d}(\mathbf{r}_{ee}) = \frac{1}{(2\pi\lambda^2)^{3/2}} e^{-\mathbf{r}_{ee}^2/2\lambda^2}. \tag{12}$$

Hence the distribution of the scalar end–to–end distance $r_{ee} = |\mathbf{r}_{ee}|$ is

$$P_{1d}(r_{ee}) = 4\pi r_{ee}^2 \frac{1}{(2\pi\lambda)^{3/2}} e^{-r_{ee}^2/2\lambda^2} \tag{13}$$

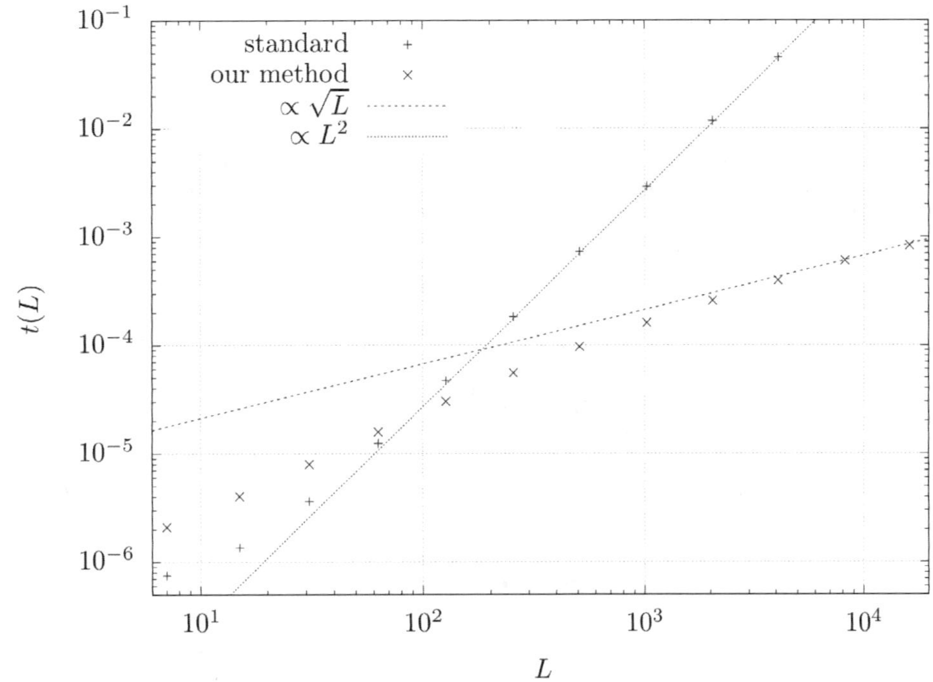

Figure 4. A comparison of the required times (in seconds) for a combination of a pivot and a bond-rotation update with a naive method and our algorithm at $T = 4$.

and it is

$$\frac{P_{1d}(r_{ee})}{r_{ee}^2} \propto e^{-r_{ee}^2/2\lambda^2}. \tag{14}$$

This means that there is a linear relation between $\ln\left(P_{1d}(r_{ee})/r_{ee}^2\right)$ and r_{ee}^2 for a GRW and consequently that a walk is behaving like a GRW if the second-order contribution a_2 of a quadratic fit,

$$\ln\left(P_{1d}(r_{ee})/r_{ee}^2\right) \approx a_0 + a_1 r_{ee}^2 + a_2 (r_{ee}^2)^2, \tag{15}$$

of this relation vanishes. We can, therefore, use the respective temperature as a finite-size estimate $T_\Theta(L)$ for the temperature of the Θ-point. The relevant data for a chain with $L + 1 = 4096$ is shown in Fig. 5.

It is neither entirely clear which analytical form the finite-size Θ-temperatures should take nor how the method of determining $T_\Theta(L)$ influences which corrections to scaling apply. Pragmatically, we have chosen from among multiple plausible approaches listed in Ref. [6], the one that works best,

$$\frac{1}{T_\Theta(L)} = \frac{1}{T_\Theta} + \frac{c_1}{\sqrt{L}} + \frac{c_2}{L}. \tag{16}$$

The fit (shown in Fig. 6) done for sizes $L > 100$ neatly describes the obtained values and provides $T_\Theta = 3.916(3)$, $c_1 = 0.26(1)$, and $c_2 = 10.2(2)$. However, as can be seen in the inset for large L the data shows a slightly more shallow slope than the fit. Using only the four largest sizes $L > 2000$ we obtain $T_\Theta = 3.907(2)$.

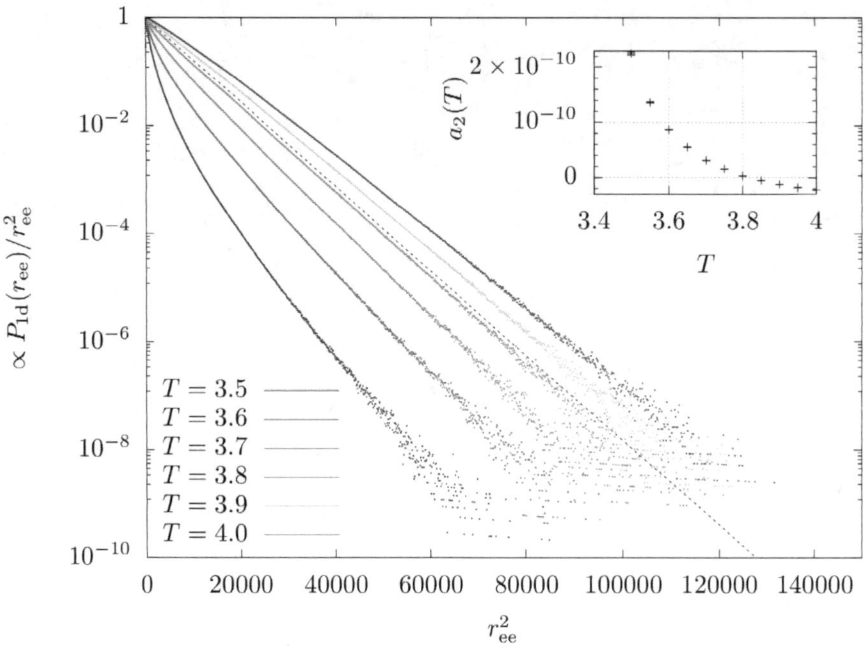

Figure 5. Distributions of the scalar end-to-end distance divided by r_{ee}^2 for different temperatures and $L + 1 = 2^{12}$. The curves are normalized to unity at $r_{ee}^2 = 0$. The dashed line is a guide to the eye. Inset: Fitting the quadratic ansatz from Eq. (15) leads to second-order coefficients $a_2(T)$ that change sign close to $T = 3.82$.

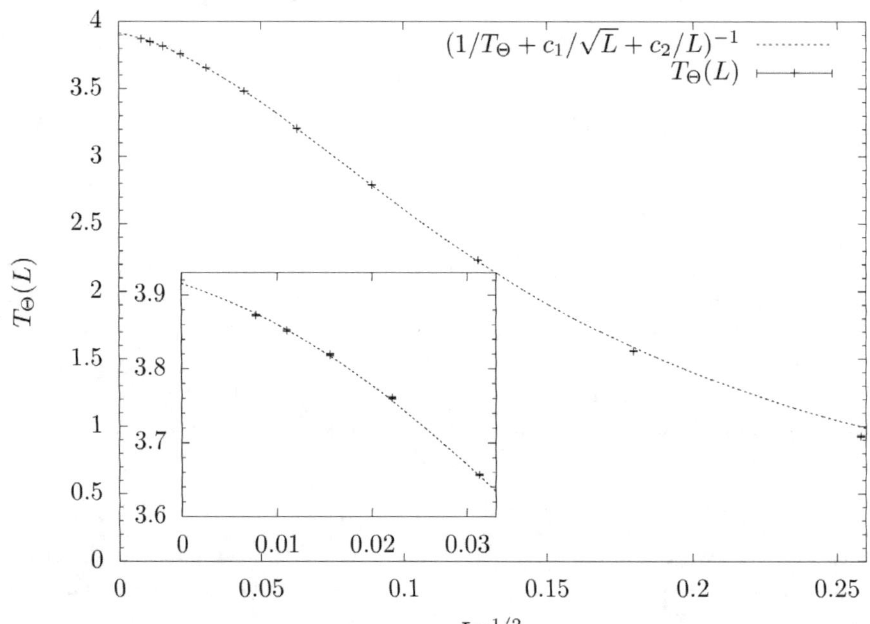

Figure 6. Finite-size estimates for the Θ-temperature. The three rightmost data points ($L < 100$) were excluded from the fit. The inset shows a close-up of the main plot for large L.

7. Conclusion

In this study we have shown how the Metropolis algorithm can be executed in a parsimonious manner. It improves the performance of the method if the change in energy associated with a conformational update can be estimated sufficiently faster than exactly calculated. When applied to a polymer with long-range 12–6 Lennard-Jones at a temperature close to the Θ-point we find that asymptotically the time required for single updates scales with $\mathcal{O}(\sqrt{L})$ or better which is a substantial improvement over the quadratic scaling required by standard methods. We were able to simulate chains with $L + 1 = 16384$ repeat units, i.e., a Hamiltonian with about 10^8 interactions.

Using the distributions of the end–to–end distance we estimated the Θ–temperature as a function of chain length. A simple finite-size scaling analysis suggests that in the infinite chain limit $T_\Theta \approx 3.91$.

Acknowledgement

The project was funded by Deutsche Forschungsgemeinschaft (DFG, German Research Foundation) under Grant No. 189 853 844–SFB/TRR 102 (project B04). We thank Nathan Clisby for helpful remarks.

References

[1] Grassberger P and Hegger R 1995 *J. Chem. Phys.* **102** 6881
[2] Frauenkorn H and Grassberger P 1997 *J. Chem. Phys.* **107** 9599
[3] Graessley W W and Hayward R C 1999 *Macromolecules* **32** 3510
[4] Rampf F, Paul W and Binder K 2006 *J. Polym. Sci. Part* B: *Polym. Phys.* **44** 2542
[5] Li B, Sun Z, An L and Wang Z 2016 *Sci. China: Chem.* **59** 619
[6] Vogel T, Bachmann M and Janke W 2007 *Phys. Rev.* E **76** 061803
[7] Metropolis N, Rosenbluth A W, Rosenbluth M N, Teller A H and Teller E 1953 *J. Chem. Phys.* **21** 1087
[8] Clisby N 2010 *J. Stat. Phys.* **140** 349
[9] Schnabel S and Janke W 2020 *Comput. Phys. Commun.* **256** 107414

Simulating Met-Enkephalin With Population Annealing Molecular Dynamics

Henrik Christiansen[1], Martin Weigel[2], Wolfhard Janke[1]
[1]Institut für Theoretische Physik, Universität Leipzig, IPF 231101, 04081 Leipzig, Germany
[2]Centre for Fluid and Complex Systems, Coventry University, Coventry, CV1 5FB, England

E-mail:
henrik.christiansen@itp.uni-leipzig.de
martin.weigel@coventry.ac.uk
wolfhard.janke@itp.uni-leipzig.de

Abstract. Met-enkephalin, one of the smallest opiate peptides and an important neurotransmitter, is a widely used benchmarking problem in the field of molecular simulation. Through its range of possible low-temperature conformations separated by free-energy barriers it was previously found to be hard to thermalize using straight canonical molecular dynamics simulations. Here, we demonstrate how one can use the recently proposed population annealing molecular dynamics scheme to overcome these difficulties. We show how the use of multi-histogram reweighting allows one to accurately estimate the density of states of the system and hence derive estimates such as the potential energy as quasi continuous functions of temperature. We further investigate the free-energy surface as a function of end-to-end distance and radius-of-gyration and observe two distinct basins of attraction.

1. Introduction

The simulation of biomolecular systems with molecular dynamics and related techniques is now a field of vast importance, including applications of paramount fundamental relevance such as protein folding [1] as well as crucial practical challenges such as those encountered, for instance, in drug discovery [2]. The dominant difficulty in this context results from the slow relaxation of such systems in the standard (micro-)canonical dynamics, that typically is a consequence of the presence of several minima in the free-energy landscape [3]. In the past 35 years or so, an array of tools designed to take on this arduous task have been developed. In Monte Carlo, the concept of *generalized ensembles* has led to a number of successful approaches, including simulated tempering [4], parallel tempering [5, 6], and multicanonical simulations [7]. Parallel tempering or replica exchange uses several copies of the system run at different temperatures, and conformation exchanges of copies at nearby temperatures allow for the efficient exploration of the free-energy landscape through the escape to a regime of fast relaxation. While initially proposed for Monte Carlo, this successful meta-algorithm was subsequently also adapted to molecular dynamics [8, 9]. For the latter, a number of alternative strategies for accelerated simulations have been proposed, including so-called "accelerated" simulations [10] as well as metadynamics [11]. These, and related techniques, are based on the idea of adaptively biasing the weight function to gradually overcome energy barriers and enable the sampling of a wide range of the relevant reaction coordinates. In this sense, these techniques are akin to the multicanonical [7] and Wang-Landau [12] methods in the world of Monte Carlo simulations.

While these methods can significantly extend the degree to which the free-energy landscape is being explored and accelerate the convergence to equilibrium, simulations of relevant biopolymers still frequently require very substantial computational resources. The significant growth in the number of cores in available (super-)computers is unfortunately only of limited utility to simulations using these techniques, as exploration of phase space and convergence to equilibrium are intimately linked to the number of time steps, and there is only limited scope for speeding up each step through (parallel) task splitting [13]. A natural way of using the available parallel resources consists of running many short simulations independently and combining the resulting statistics to improve the degree of sampling of the relevant states [14]. While this can work reasonably well, it does not remove the bottleneck of equilibration as each simulation conducted in parallel must at least run for the time required to thermalize. The popular use of Markov state models describing transitions between the valleys [15, 16] can only lead to trustworthy results if a reliable picture of the relevant states has already been deduced from direct simulations before.

In the present contribution, instead, we demonstrate how a recently introduced massively parallel approach to molecular dynamics (MD) simulations, population annealing molecular dynamics (PAMD), can be employed to utilize practically any amount of parallel resources available to improve the sampling in systems that are found hard to thermalize [17, 18]. Population annealing, that was first introduced in the context of Monte Carlo simulations [19, 20, 21, 22], uses a large population of identical system copies (replicas) which are successively cooled down starting from a thermalized population at high temperatures where equilibration is straightforward. At each cooling step, the population is resampled, favoring copies well adapted to the lower temperature. This process is reminiscent of genetic algorithms [23] where a population is evolved according to its fitness. For each new generation mutations occur at a prescribed rate, i.e., random changes in the microscopic variables. Subsequently, the new resulting fitness is calculated. According to some determined threshold, conformations are replicated or pruned. Genetic algorithms are hugely successful as optimization algorithms, i.e., in finding ground-state configurations [24, 25]. However, no thermodynamic information can be extracted. On the other hand, PAMD provides an equilibrium sample at each temperature considered, and it can hence be seen as an equilibrium variant of a combination of simulated annealing [26] and a genetic algorithm.

2. Model and Method

The system studied here is Met-enkephalin, a penta-peptide with amino-acid sequence Tyr-Gly-Gly-Phe-Met, that occurs in many organisms. It inhibits neurotransmitter release upon activation of the appropriate opioid receptor, and so plays a central role in pain regulation [27]. It is said to inhibit tumor growth and metastasis [28]. For the simulation, we cap the ends with a methyl and an acetyl group, respectively, resulting in a total size of 84 atoms. The interactions were modeled by the AMBER force-field ff94 [29].

While previous simulation studies of this system mostly used the replica exchange method for ensuring that equilibration is achieved, we here employ PAMD simulations to study the system [17]. In particular, this involves the following simulation steps:

(i) Set up an equilibrium ensemble of R independent copies of the peptide at some high temperature T_0.

(ii) Choose the next temperature $T_i < T_{i-1}$ from a pre-defined sequence.

(iii) Resample the population of systems to the new temperature T_i by replicating each copy a number of times proportional to the relative Boltzmann weight $\tau_j \propto e^{-(1/k_BT_i - 1/k_BT_{i-1})E_j}$,

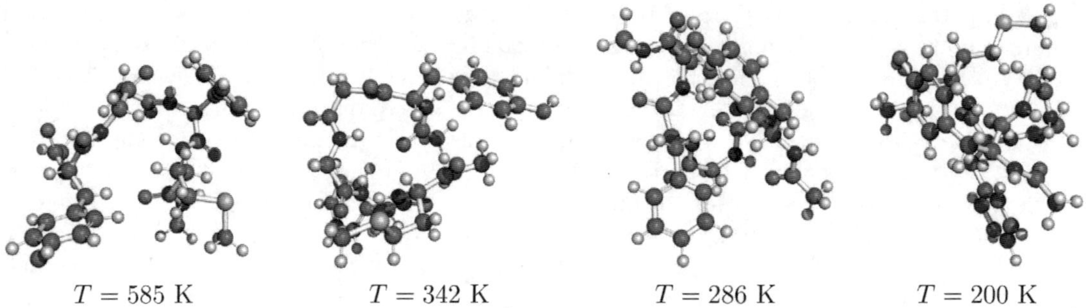

Figure 1. Snapshots of Met-enkephalin at different temperatures, decreasing from left to right. The following color-code applies: pink – carbon, white – hydrogen, red – oxygen, blue – nitrogen, yellow – sulfur.

where E_j is the potential energy of the jth replica. The momenta are simply adjusted to the new temperature by rescaling $p_k \to \sqrt{\frac{T_i}{T_{i-1}}} p_k$.

(iv) Update each replica with θ simulation steps of the underlying MD algorithm.

(v) Calculate observables $\mathcal{O}$ at temperature T_i as population averages.

(vi) Goto step (ii) until T_i reaches or falls below the target temperature T_N.

In parallel tempering, the choice of a suitable temperature protocol is often a challenging task [30, 31]. For PAMD, on the other hand, we have shown in Ref. [18] that an appropriate temperature set may be found "on the fly" by demanding a constant energy histogram overlap without the need of conducting preliminary simulations. For the present study we adopt the previously used temperature scheme consisting of simulations at T_i = 700, 585, 489, 409, 342, 286, 239, 200 K. This corresponds to a constant energy histogram overlap of about 30% [18]. Due to the resampling that involves making copies of some population members, it is crucial to employ a stochastic thermostat in PAMD simulations, as otherwise the copies would eternally follow the same trajectory, thus compromising the quality of the statistical sampling [17]. Here we use a Langevin thermostat with 0.5 fs integration step and friction coefficient $\gamma = 1$/ps. For our simulation we ran 10^4 replica with a total simulation time of 200 ns. MD steps amounting to 25 ns were used to initialize the population at the highest temperature, followed by 21.875 ps of evolution (or 4375 updates) performed at each temperature step. For the molecular dynamics part, which dominates the computational effort, we relied on the package OpenMM [32]. Overall, our simulations show an excellent parallel efficiency of over 85% using 500 cores, far superior to what normally can be achieved in parallel tempering.

3. Results

To convey an overall impression of the behavior of Met-enkephalin, we present typical snapshots of the peptide in Fig. 1. It is clearly seen that compared to the initial configuration at the highest temperature (left), the conformations obtained by population annealing at lower temperatures (to the right) are significantly more compact.

Thermodynamic properties at the temperatures included in the simulation protocol can be extracted from regular population averages. Beyond that, effectively continuous estimates of observables can be derived from the density of states $\Omega(E)$ that can be obtained using the weighted-histogram analysis method (WHAM) [33, 34] applied to the potential energy E. Details of the application to population annealing can be found in Ref. [35]. The result of this analysis for Met-enkephalin is shown in Fig. 2(a). Note that due to the continuous nature of the energy

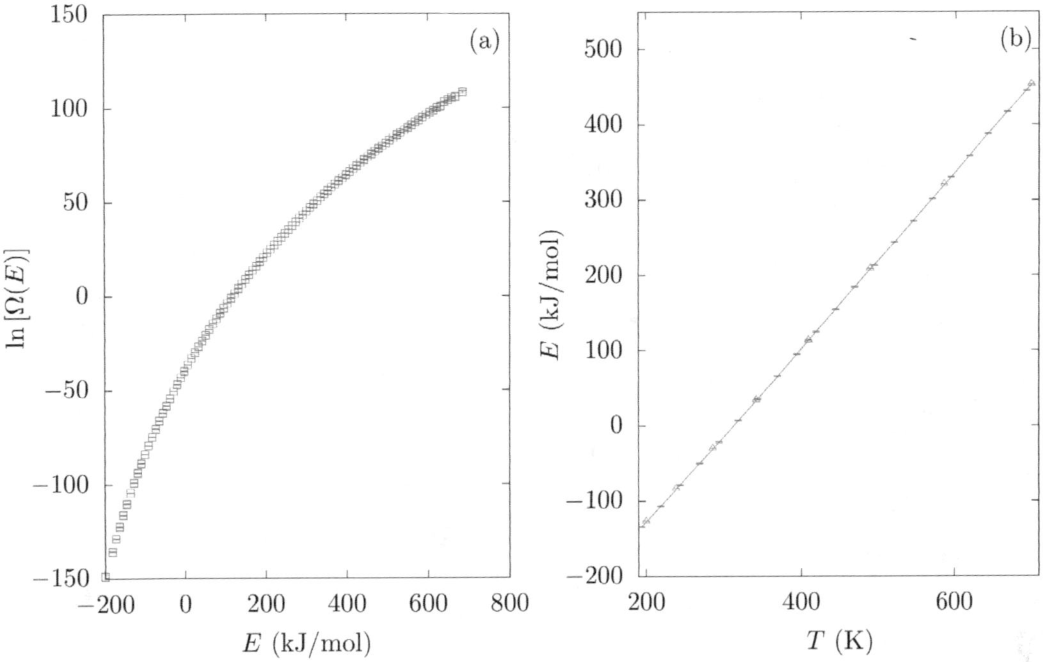

Figure 2. (a) Logarithm of the density of states for Met-enkephalin obtained from our PAMD simulation using WHAM [33, 34, 35]. (b) (Potential) energy E as derived from the density of states of (a) as a function of temperature T. The green triangles indicate population averages at the actual simulation temperatures.

functional one needs to use binning to apply WHAM. Here, we used 200 equally spaced bins in the range from the lowest energy observed to the highest. (The binning is only applied after the simulation, in the post-processing of data.) Using this result for the density of states, it is then possible to extract observable estimates at any temperature such as

$$\langle E \rangle(T) = \frac{\sum_E E\Omega(E)e^{-E/k_BT}}{\sum_E \Omega(E)e^{-E/k_BT}}. \quad (1)$$

The resulting plot is shown in Fig. 2(b) for temperatures in the simulation range of 200 K $< T <$ 700 K.

The spatial structure of the peptide can be characterized using quantitative measures such as the end-to-end distance,

$$R_{\rm ee} = |\mathbf{R}_N - \mathbf{R}_1|, \quad (2)$$

where $\mathbf{R}_N$ is the position of the last atom of the peptide and $\mathbf{R}_1$ that of the first. For a peptide it is not particularly useful to apply this definition directly and, instead, we rather use the distance between the two outermost carbon atoms. While it is known that this quantity can be misleading, e.g., when "by chance" the start and end monomer in an otherwise extended conformation are very close to each other, it may provide useful insight when directly compared to other physical quantities. A more robust indicator of the average size is the (squared) radius-of-gyration,

$$R_{\rm g}^2 = \frac{\sum_{i=1}^N m_i(\mathbf{R}_i - \mathbf{R}_{\rm com})^2}{\sum_{i=1}^N m_i}, \quad (3)$$

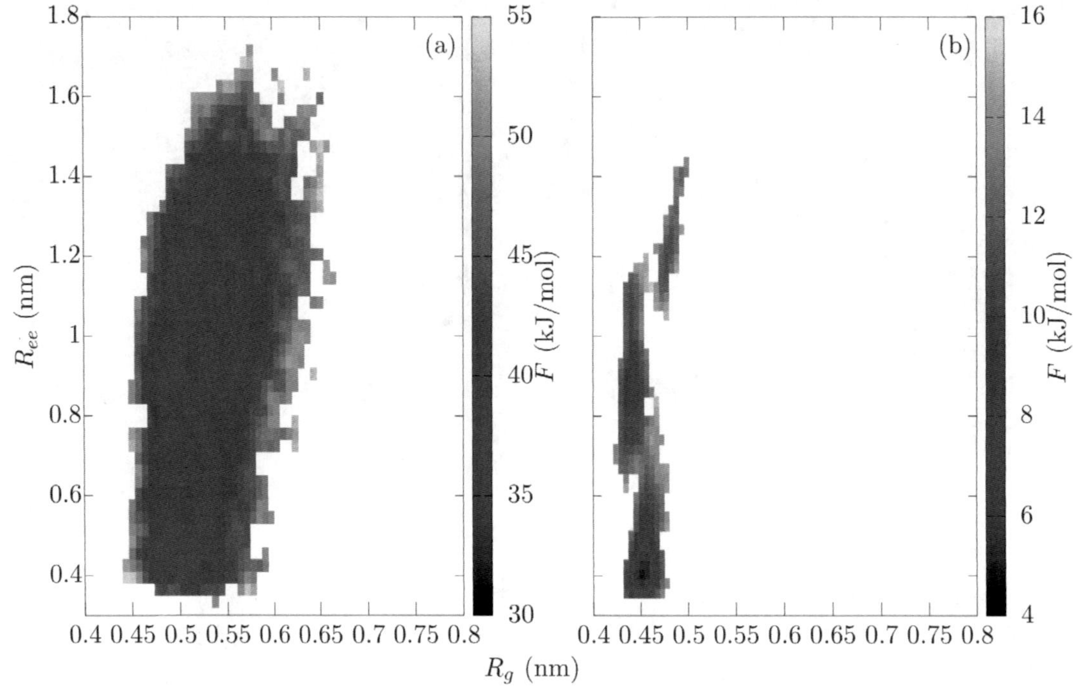

Figure 3. Plots of the end-to-end distance versus the radius-of-gyration for our simulation of Met-enkephalin at (a) $T = 700$ K and (b) $T = 200$ K.

where the sum from $i = 1$ to N runs over all atoms and $\mathbf{R}_{\mathrm{com}} = \sum_i m_i \mathbf{R}_i / \sum_i m_i$ is the center-of-mass position of the peptide. This observable thus averages the squared distance of each atom weighted by its mass to the center-of-mass and is hence a well defined measure for the current extension of the peptide.

While it can already be useful to consider averages and potentially higher moments of quantities such as R_{ee} and R_{g}, it is even more instructive to consider their full probability distributions. In general, define the probability density of an observable $\mathcal{O}$ as

$$P(\mathcal{O}) \sim \iint \mathrm{d}\mathbf{q}\mathrm{d}\mathbf{p}\, P(\mathbf{q},\mathbf{p})\delta\left[\mathcal{O}(\mathbf{q},\mathbf{p}) - \mathcal{O}\right], \tag{4}$$

where $\mathbf{q}$ are the positions and $\mathbf{p}$ the momenta of the atoms. It is then sometimes useful to analyze the *free-energy surface*

$$F(\mathcal{O}) = -k_B T \ln\left[P(\mathcal{O})\right] + c. \tag{5}$$

The constant c is arbitrary and chosen to be zero here. This signifies the observation that only free-energy differences are significant, as the normalization is unknown from simulations.

As an illustration, we plot in Fig. 3 the free-energy surfaces of the end-to-end distance versus the radius-of-gyration for (a) the highest temperature of 700 K and (b) the lowest temperature of 200 K. As expected, both the end-to-end distance and radius-of-gyration are significantly smaller for lower temperatures, where the peptide is in a more compact conformation. At $T = 200$ K the free-energy landscape splits in two minima separated by a region in which no conformations are observed. This signifies two different folding states, which using canonical simulations would not have been observable in a single run.

4. Conclusion

We have described population annealing for molecular dynamics and performed simulations for the penta-peptide Met-enkephalin, illustrating the utility of this approach for efficiently thermalizing the system and estimating thermodynamic quantities. With the help of the weighted-histogram analysis method it is possible to get reliable estimates of the density of states that in turn enable the derivation of quasi continuous estimates of observables as a function of temperature. This peptide has several free-energy minima, which we present by plotting the free-energy surface of the end-to-end distance versus the radius-of-gyration. Here, at least two basins of attractions are observed, indicating the presence of a free-energy barrier at $T = 200$ K. In the population simulation method employed here, at low temperatures different basins are occupied by replicas according to their statistical weight, thus ensuring a fair sampling of the full free-energy landscape while allowing the efficient use of massively parallel computational resources.

Acknowledgments

This project was funded by the Deutsche Forschungsgemeinschaft (DFG, German Research Foundation) under Grant No. 189 853 844 – SFB/TRR 102 (project B04), and further supported by the Deutsch-Französische Hochschule (DFH-UFA) through the Doctoral College "$\mathbb{L}^4$" under Grant No. CDFA-02-07, the EU Marie Curie IRSES network DIONICOS under Grant No. PIRSES-GA-2013-612707, and the Leipzig Graduate School of Natural Sciences "BuildMoNa".

References

[1] Scheraga H A, Khalili M and Liwo A 2007 *Annu. Rev. Phys. Chem.* **58** 57–83
[2] Kitchen D B, Decornez H, Furr J R and Bajorath J 2004 *Nat. Rev. Drug Discov.* **3** 935–949
[3] Janke W (ed) 2007 *Rugged Free Energy Landscapes — Common Computational Approaches to Spin Glasses, Structural Glasses and Biological Macromolecules* (*Lect. Notes Phys.* vol 736) (Berlin: Springer)
[4] Marinari E and Parisi G 1992 *Europhys. Lett.* **19** 451–458
[5] Geyer C J 1991 Markov chain Monte Carlo maximum likelihood *Computing Science and Statistics: Proceedings of the 23rd Symposium on the Interface* (New York: American Statistical Association) pp 156–163
[6] Hukushima K and Nemoto K 1996 *J. Phys. Soc. Jpn.* **65** 1604–1608
[7] Berg B A and Neuhaus T 1992 *Phys. Rev. Lett.* **68** 9–12
[8] Hansmann U H E 1997 *Chem. Phys. Lett.* **281** 140–150
[9] Sugita Y and Okamoto Y 1999 *Chem. Phys. Lett.* **314** 141–151
[10] Hamelberg D, Mongan J and McCammon J A 2004 *J. Chem. Phys.* **120** 11919
[11] Laio A and Parrinello M 2002 *PNAS* **99** 12562–12566
[12] Wang F and Landau D P 2001 *Phys. Rev. Lett.* **86** 2050–2053
[13] Weigel M 2018 Monte Carlo methods for massively parallel computers *Order, Disorder and Criticality* vol 5 ed Holovatch Y (Singapore: World Scientific) pp 271–340
[14] Gross J, Zierenberg J, Weigel M and Janke W 2018 *Comput. Phys. Commun.* **224** 387–395
[15] Chodera J D, Singhal N, Pande V S, Dill K A and Swope W C 2007 *J. Chem. Phys.* **126** 155101
[16] Pande V S, Beauchamp K and Bowman G R 2010 *Methods* **52** 99–105
[17] Christiansen H, Weigel M and Janke W 2019 *Phys. Rev. Lett.* **122** 060602
[18] Christiansen H, Weigel M and Janke W 2019 *J. Phys. Conf. Ser.* **1163** 012074
[19] Iba Y 2001 *Trans. Jpn. Soc. Artif. Intell.* **16** 279–286
[20] Hukushima K and Iba Y 2003 *AIP Conf. Proc.* **690** 200–206
[21] Machta J 2010 *Phys. Rev. E* **82** 026704
[22] Barash L Y, Weigel M, Borovský M, Janke W and Shchur L N 2017 *Comput. Phys. Commun.* **220** 341–350
[23] Holland J H *et al.* 1992 *Adaptation in Natural and Artificial Systems: An Introductory Analysis with Applications to Biology, Control, and Artificial Intelligence* (MIT Press, Cambridge, USA)
[24] Deaven D M and Ho K M 1995 *Phys. Rev. Lett.* **75** 288–291
[25] Beasley J E and Chu P C 1996 *Eur. J. Oper. Res.* **94** 392–404
[26] Kirkpatrick S, Gelatt C D and Vecchi M P 1983 *Science* **220** 671–680
[27] Hughes J, Smith T, Kosterlitz H, Fothergill L A, Morgan B and Morris H 1975 *Nature* **258** 577–579

[28] Kuniyasu H, Luo Y, Fujii K, Sasahira T, Moriwaka Y, Tatsumoto N, Sasaki T, Yamashita Y and Ohmori H 2010 *Gut* **59** 348–356
[29] Cornell W D, Cieplak P, Bayly C I, Gould I R, Merz K M, Ferguson D M, Spellmeyer D C, Fox T, Caldwell J W and Kollman P A 1995 *J. Am. Chem. Soc.* **117** 5179–5197
[30] Katzgraber H G, Trebst S, Huse D A and Troyer M 2006 *J. Stat. Mech.: Theory and Exp.* **2006** P03018
[31] Bittner E, Nussbaumer A and Janke W 2008 *Phys. Rev. Lett.* **101** 130603
[32] Eastman P, Swails J, Chodera J D, McGibbon R T, Zhao Y, Beauchamp K A, Wang L P, Simmonett A C, Harrigan M P, Stern C D *et al.* 2017 *PLOS Comput. Biol.* **13** e1005659
[33] Ferrenberg A M and Swendsen R H 1989 *Phys. Rev. Lett.* **63** 1195–1198
[34] Kumar S, Rosenberg J M, Bouzida D, Swendsen R H and Kollman P A 1992 *J. Comput. Chem.* **13** 1011–1021
[35] Barash L, Marshall J, Weigel M and Hen I 2019 *New. J. Phys.* **21** 073065

Journal of Physics: Conference Series **2241** (2022) 012007 doi:10.1088/1742-6596/2241/1/012007

GPU molecular dynamics: Algorithms and performance

D C Rapaport

Department of Physics, Bar-Ilan University, Ramat-Gan 52900, Israel

E-mail: rapaport@mail.biu.ac.il

Abstract.
A previous study of MD algorithms designed for GPU use is extended to cover more recent developments in GPU architecture. Algorithm modifications are described, togther with extensions to more complex systems. New measurements include the effects of increased parallelism on GPU performance, as well as comparisons with multiple-core CPUs using multitasking based on CPU threads and message passing. The results show that the GPU retains a significant performance advantage.

1. Introduction

Improved computer performance is increasingly dependent on parallelism, a consequence of clock limits for current processor technology. Automating simpler forms of software parallelization is within the capability of modern compilers, but manual software redesign is required whenever the algorithm logic is incompatible with the processor architecture. This was true for the earliest vector and parallel supercomputers, and is equally so for modern hardware in which the GPU, or graphic processing unit, a device whose name reflects its original purpose, is the key component.

While the physical phenomena modeled by MD (molecular dynamics) simulation [1] have a major parallel component, in which particles (atoms or molecules) independently 'evaluate' the combined forces exerted on them by their surroundings, MD computations are traditionally implemented in serial form, namely, at each time step there are processing loops that iterate over all particles or particle-pairs. The availability of large-scale data parallelism eliminates the need for most of these explicit loops. As examples, the overall force computation for each particle can be carried out completely independently and in parallel (provided Newton's third law is not invoked), while the trivial sum to evaluate the total energy is inherently serial and therefore requires the use of efficient, but more complex, parallel reduction techniques.

The present paper considers MD algorithm implementation on a more advanced GPU than considered previously [2]. Measured speedups are due not only to the substantially increased hardware parallelism but also to architectural enhancements. In addition to the comparison between GPU performance and a single-threaded CPU application, comparisons have been made with alternative approaches employing multicore CPU parallelism based on CPU (or Unix) threads and message-passing. Factors associated with the algorithms and their implementations that influence GPU performance are also investigated.

2. Background and methodology

2.1. GPU hardware

The GPU is a highly parallel coprocessor, with its own private memory, that operates separately from the CPU but under its control. GPUs provide extremely fine-grained data parallelism, unlike a typical CPU with at most a few processing cores. Data can be exchanged between CPU and GPU, although for efficiency this should be kept to a minimum. GPU hardware, and the associated software, are in a state of continuous development, but have already reached a level of maturity where GPUs are the key components in many of the most powerful supercomputers currently available, while at the same time also available packaged as affordable commodity products, principally plugin graphics cards and chipsets. This suggests that GPU-based computing is destined for a long life, justifying the investment of effort in designing algorithms tuned for optimal GPU performance (product lifetime is a factor in determining the overall cost-effectiveness of novel hardware for general use). Current GPU complexity tends to require software design by trial and error, with improvements over successive hardware generations providing more flexibility that can simplify the task.

GPUs are assembled from a set of multiple-core processors and eschew many of the more complex features of conventional CPUs; the performance gains are achieved mainly by parallelism rather than the more sophisticated CPU instruction handling. To use NVIDIA nomenclature [3], a GPU device consists of multiple streaming multiprocessors (SMs), each consisting of (typically) 128 cores. A fully-configured GPU contains up to several thousand cores. Each core executes a single instruction thread, with threads grouped into blocks that execute as multiple 32-thread warps on each SM. While the threads in each block can be synchronized and can exchange data through shared memory, separate blocks are processed independently and there is no direct communication between them, with block synchronization typically occuring at kernel (defined below) completion. Other, more subtle, details associated with thread scheduling are of more specialized interest [3, 4].

The GPU includes onboard global memory; since the access latency is relatively long, measured in hundreds of clock cycles, the ability to coalesce memory requests improves read/write speeds substantially. In its simplest form, coalescence requires that blocks of threads access aligned, consecutive memory locations. Memory latency can also be hidden when there are multiple thread warps awaiting processing. Since blocks access global memory completely independently, synchronization between blocks within a kernel is nontrivial and requires special effort. Multiple caches can improve noncoalesced memory usage, a feature that varies between GPU generations. Global memory is also accessible by the host. Threads within the same block can also access much faster read/write shared memory that enables exchange of data (there are possible bank conflicts that can slow access, as well as the need for thread synchronization to prevent indeterminate race conditions). There is also a fast read-only constant memory that the host can write into. Finally, there is a set of local registers private to each thread. Further details of this complex memory scenario appear in [3, 5].

Function (or subroutine) calls are executed as GPU kernels. Each kernel call from the CPU specifies the size of the thread block and the total number of blocks. Multiple thread blocks run concurrently in an unspecified order determined by a scheduler. Explicit thread synchronization is available to ensure that threads reach a particular way-point in their execution, e.g., prior to exchanging data, although care is required to avoid deadlocks and not degrade performance. Software development typically involves CUDA C (or C++), a language extension that greatly simplifies GPU programming.

2.2. MD algorithms

Algorithms designed to utilize massive parallelism introduce novel issues. Some computations are readily converted from serial to parallel form, especially when independent threads can replace

Journal of Physics: Conference Series **2241** (2022) 012007 doi:10.1088/1742-6596/2241/1/012007

iterated operations that have no mutual effect. Others require cooperation between threads because data must be shared, such as the reduction-type operation involved in summing an array, and in more complex tasks such as sorting; these require completely different techniques that are inefficient serially but optimal when parallelized. Some changes can be handled using standardized approaches (e.g., reduction and sorting), while others need special treatment, notably the MD tabulation of atom neighbors using the neighbor matrix (below), that replaces the more familiar neighbor list due to its unsuitability for parallel use.

The three kinds of MD programs [1] that are considered here, aside from the standard serial version, are those based on job subdivision using (a) multiple CPU threads, (b) coarse-grained parallelism in which several CPU tasks, with message-passing communication, are responsible for subregions of the system, and (c) fine-grained parallelism with one GPU thread per atom. Each entails changes from the serial approach; in terms of effort to adapt the programs, CPU threads are the easiest, message-passing requires extra bookkeeping, while the GPU-based approach requires major reformulation resembling that used in vectorization [6].

The GPU algorithm uses a matrix array for cell occupancy whose filling and occupancy counting require 'atomic' update operations (in this context 'atomic' means an uninterruptible sequence of operations that are guaranteed exclusive memory access). This matrix is then accessed in the orthogonal direction to obtain the layers that contain the sets of interaction neighbors of each atom; here there are also different ways of organizing the computation, with the method described here being efficient and fully scalable.

2.3. Neighbor matrix construction and usage

The algorithm details, with an emphasis on the differences between the CPU and GPU approaches, are described in [2]. A brief summary of the GPU version of the method follows to allow subsequent performance measurements to reference the steps. Practically all the computation is carried out on the GPU, with minimal CPU involvement. Additional details of the MD computation are covered in [2] and more general considerations appear in [1].

In all cases, loops over atoms in the serial CPU program, or the outer loops when nested, are replaced by GPU threads, one per atom. Maximal parallelism is achieved, with performance scaling limited only by hardware capability, primarily the GPU core count and clock speed, and the memory bandwidth. Neighbor matrix construction is carried out at intervals of several time steps:

(N1) Assign atoms to cells, based on position; atom i is in cell c_i, with multiple atoms allowed per cell.

(N2) Assign atoms in each cell to layers, where layer l includes the lth members of each cell; atom i is in layer l_i. Each cell c requires a cell occupancy counter k_c, and due to multiple occupancy, incrementing k_{c_i} requires an 'atomic' operation.

(N3) Determine the number of layers required, N_l, by finding the maximum of k_c using a reduction operation.

(N4) Build the cell-layer occupancy matrix H by setting $H_{c_i,l_i} = i$ for each atom i; the row and column indices of $H_{c,l}$, specify the cell ($c \leq N_c$) and layer ($l \leq N_l$).

(N5) Construct the neighbor matrix W; for each atom i there are two nested loops to access the neighbors $i' = H_{c,l}$, first over the neighboring cells c of c_i, and then over layers $l \leq k_c$. The column indices i of $W_{m,i}$ correspond to atoms; each row m specifies the atoms' mth neighbors (unordered). The count of i's neighbors is accumulated in m_i.

Force evaluation, at each time step:

(F1) For each atom i accumulate the total force $\vec{f_i}$ and (optionally) interaction energy u_i by considering the subset of atoms $i' = W_{m,i}$ for $m < m_i$ that lie within interaction range.

(F2) Sum the individual u_i to obtain the total interaction energy U (actually $2U$) using a reduction operation (optional).

Comments on the computations follow, including ways that the GPU architecture affects the algorithm.

There can be a performance penalty for 'atomic' GPU operations; this depends on the GPU architecture. In [2] better performance was obtained when the l_i evaluation (N2) was carried out on the CPU, even after allowing for the additional data transfers. With more recent GPUs this is no longer the case. Note that situations requiring 'atomic' operations also lead to irreproducible floating-point results since the order of layer occupants in each cell is indeterminate; changing the order in which atoms are processed can alter the lowest-order mantissa bits because the arithmetic is nonassociative, but for MD this sensitivity is masked by the inherently chaotic nature of the atomic trajectories [1].

Efficient evaluation of global quantities on the GPU requires (nontrivial) parallel reduction operations; examples include the total interaction and kinetic energies, and the maximal atom velocity needed to determine when the neighbor matrix needs rebuilding. Reduction operations rely on efficient use of shared memory and follow a standard hierarchical pattern that maximizes the work carried out in parallel [7]. In (N3) and (F2) the reductions are carried out on the GPU, producing a single result per thread block, with a minor completion step on the CPU after a small data transfer.

The final reductions, involving one result per thread block, can also be carried out on the GPU. Since the number of thread blocks is much less than the number of atoms, by two orders of magnitude, and the final result is usually required by the CPU, the performance benefit of the extra GPU work is limited, but it does allow greater task separation between GPU and CPU. The GPU implementation [3] uses an 'atomic'ally updated counter in global memory to record the number of blocks that complete their work (since thread blocks are independent this is the most a block can know about mutual progress) as well as a 'memory fence' (lockout) function to ensure that write operations are completed before proceeding. The block that finds itself to be the last to finish can then finalize the reduction task, using its threads to process the data output by all blocks (itself included) to global memory, yielding a single result.

The layout of W, assuming the matrix to be stored by rows, allows the identities of all mth neighbors to occupy successive memory locations; this permits coalesced access by threads processing individual atoms in (F1). Atom pairs appear twice in W and are considered twice in (F1) because Newton's third law is not invoked; instead of using each atom pair (i, i') to update the (equal but opposite) forces acting on both atoms, $\vec{f_i}$ and $\vec{f_{i'}}$, where atoms i are accessed sequentially but atoms i' essentially at random, only $\vec{f_i}$ is updated. This allows more efficient coalesced memory access for atoms i that compensates for the extra computation.

2.4. Software design

A number of considerations influenced the overall software design.

(1) The variables describing the state of each atom, namely its position, velocity and acceleration, are three-component vectors. Representing them on the GPU as `float4` quantities allows efficient memory access, more than justifying the extra storage; the unused fourth component can hold other information, such as the interaction energy during the force calculation. Use of an array of structures that combine different kinds of state data for each atom, such as position and velocity, is a 'recommended' form of data organization; unfortunately, it is incompatible with efficient GPU operation since it inhibits coalesced memory access and degrades performance.

(2) Memory allocation on the GPU is requested by CPU function calls. The GPU memory pointers returned must later be supplied to the GPU. This could be done by including them as arguments to GPU kernel calls. However, since the computations involve several arrays as well as other parameters, all this unvarying information is placed in a small data structure and copied to a region of GPU storage called the constant memory where it can be efficiently accessed by

all threads; this eliminates the need for long argument lists that would otherwise accompany each kernel call.

(3) Essentially all computation is done on the GPU to avoid relatively slow CPU-GPU data transfers. After constructing the initial state on the CPU, almost all the state data exists only on the GPU. Results of measurements are returned to the CPU when required. If visualization is employed, atom coordinates must be made available by the GPU for updating the imagery; if a single GPU handles both computation and graphics, even this data transfer can be avoided. Checkpoint and restart would also require transfers of most of the state data.

(4) An exception to using the GPU for everything is the infrequent data rearrangement (sorting) required to reorder atoms for efficient memory access. The problem is to pack the nonzero elements of H into a vector s of length N_a that specifies the reordered atom sequence (determined by cell membership); though trivial when carried out serially, the efficient parallel version is more complicated. Here, reordering involves copying the most recent version of H to the CPU; this is used to construct s which is returned to the GPU and used to reorder atom coordinates and velocities (using a temporary buffer, with caching aided by the fact that data is already partially ordered), following which the matrices H and W are rebuilt. These data transfers have negligible performance impact. The GPU alternative would be to apply a parallel prefix-scan [8] to the set of k_c that specify the cell occupancy; each element of the result is a cumulative sum of k_c for the preceding cells. Cells are then processed in parallel, since each thread knows where to place its cell's atom identities in s. In general, small serial tasks that require excessive effort to implement on the GPU are worth converting only if there is a clear need to avoid data transfers with the host, or if suitable, well-tuned library code is available. A similar compromise was adopted with the reduction operations used for energy (etc.) evaluation (see above).

(5) There is an advantage to having a single program source that, during the compilation phase, generates code for either just the CPU or the combined GPU/CPU environment, rather than having to maintain separate versions; conditional statements in the source file(s) control this process. The NVIDIA `nvcc` compiler [3] deals with all the GPU kernel code and then automatically calls the `gcc` C compiler to process the CPU code and the linking; for the CPU version of the program, only the C compiler is used. Debugging is simpler on the CPU, so that issues not specific to the GPU can be resolved simultaneously in both versions of the code. This approach assumes that the same algorithm is being used in both cases and the changes are principally the replacement of loops by parallel threads, as well as standardized reduction procedures, which indeed cover the most frequent changes when converting CPU code for GPU use.

3. Performance measurements

3.1. Hardware and software configuration

The NVIDIA GPU considered here is the mobile version of the P4000, based on the (recent, but not latest) Pascal architecture [5]. Detailed GPU design is subject to frequent change, with some changes affecting performance significantly, and others less so; the consequences are not always apparent from the hardware specifications. The principal feature characterizing a GPU is the core count, reflecting its parallel capability. Given the limits to Moore's law, increasing core count is the main route to faster performance; the P4000 has 1792 cores. Processor clock speed is also important; here it is 1227MHz, however, what the GPU is capable of doing in a single cycle (e.g., multiple operations per thread) varies with model. Memory access is complicated [3, 4, 9], but the overall bandwidth is important; here the value is 192 GB/s. There are numerous other contributing factors, but the nominal peak rating is 4398 s.p. (single precision) GFlop/s. While these numbers are purely theoretical, and unachievable, the difference is reflected in the actual performance. By comparison, the earlier [2] FX770M had a very modest 32 cores, 500MHz clock,

Table 1. Size dependence for soft sphere (SP) and Lennard-Jones (LJ) systems; the numbers of atoms ($N_a = {N_e}^3$) and the times per atom-step (t, in μs) are shown.

	N_e	N_a	t_{SP}	t_{LJ}
GPU	32	32768	0.00546	0.01152
	64	262144	0.00363	0.00850
	96	884736	0.00356	0.00881
	128	2097152	0.00373	0.00925
	160	4096000	0.00386	0.00898
	192	7077888	0.00386	0.00919
	224	11239424	0.00387	
	256	16777216	0.00386	
CPU	48	110592	0.134	0.558

26 GB/s bandwidth, and a peak 80 GFlop/s, so a considerable speedup can be anticipated.

Several aspects of performance will be analyzed. Comparisons of GPU speed relative to a single CPU core are considered first. These are followed by tests comparing the increased CPU capability made available by using multiple CPU cores in two different ways. Factors contributing to GPU performance are considered; these can sometimes reveal aspects of behavior not obvious from the published device specifications.

The system parameters used for the tests strongly affect the results; they are the same as in [2] and are as follows (all in reduced MD units), unless otherwise specified: The soft-sphere system, denoted by SP, has density $\rho = 0.8$, temperature $T = 1$, and interaction cutoff $r_c = 2^{1/6} = 1.122$. For the LJ system the values are $\rho = 0.38$, $T = 1.2$ and $r_c = 2.5$. In both cases particle reordering occurs every 100 time steps, the shell thickness used for neighbor matrix is $\delta = 0.6$ (in [2] 0.8 was used for LJ), and the integration time step is 0.005. The initial state consists of a cubic grid of atoms of size $N_a = {N_e}^3$ that are assigned random velocities. Runs are of length 5000 time steps.

GPU measurements were carried out on a Dell Precision 7720 notebook workstation (Intel 4-core Xeon E3-1505Mv6 CPU) running Ubuntu Linux; GPU development is based on CUDA version 8. The other computer referenced in this work (for the multicore CPU comparisons) is an HP Z820 dual-CPU workstation (two 4-core Intel Xeon E5-2637v2 CPUs) running Centos Linux (with d.p. computation that is only slightly slower than s.p.); compilation uses the `gcc` compiler, run at optimization level `O3`; the CPU comparisons use the additional cores of this slightly faster processor pair.

3.2. GPU tests

A series of GPU speed measurements for the SP and LJ systems over a range of sizes, extending beyond those considered in [2], appear in Table 1. These results are for the most efficient version of the algorithm, run under optimal conditions; factors determining these conditions are discussed later.

Over most of the size range, excluding the smallest system that does not allow the GPU to reach full capacity, there is no systematic size dependence (cell size has a small effect on the results). Results for the best CPU version of the MD algorithm on a single CPU core are included for comparison. The P4000 GPU is seen to provide speedups relative to the CPU of 34.7x and 60.7x for the SP and LJ cases. There is also a 20-fold speedup over the original FX770, that lies between the increased 8x memory bandwidth and 50x core count with doubled

Table 2. Percentage of time spent in different parts of the MD calculation.

Task	SP	LJ
nebr matrix:	44.0	58.1
forces:	26.8	29.8
integration:	20.1	8.4
energy:	5.9	2.5
layer cells:	1.7	0.7
reorder:	1.4	0.4

clock speed. Running the layer-based algorithm on the CPU halves the performance (allowing the dubious claim of a further 2x GPU speedup).

Knowing where the computational effort is expended helps focus optimization attempts. Table 2 shows the time fractions devoted to the steps of the calculation; the results depend on the GPU model and the system state. Work associated with neighbor matrix construction and force evaluation accounts for the majority of the processing time, with the former, manifestly unsuited for the GPU, dominating. The neighbor matrix is typically refreshed approximately every 10 steps (see below). On the CPU the situation is very different, and the respective time fractions for SP(LJ) are 0.26(0.22) and 0.63(0.76). Another consideration is increased storage of the layer approach, although here this is not an issue; storage per atom for SP(LJ) is 224(474) bytes, and the largest runs required a total of 3-4 GBytes out of the 8 GBytes installed.

3.3. Multiple-core CPU comparisons

Since modern CPUs incorporate multiple cores, comparisons of GPU and CPU performance should include parallel tests utilizing the available cores. The two ways of decomposing a computation to use one or both CPUs in a single processing node and the multicore CPUs themselves employ (a) CPU (or Unix) threads (no relation to GPU threads) that access common data, or (b) multiple CPU processes that communicate only via OpenMPI message-passing with each maintaining its own data storage, the latter directly transferable to multiple nodes connected over a network. In either case the MD software requires adaptation, although considerably less in the former; the details appear in [1]. Table 3 summarizes tests run on the dual-CPU workstation that allowed up to 8 threads and processes.

The observed diminishing returns are due to the extra communication and computation needed to support increased parallelism. The GPU maintains its dominance, delivering 5x the speed of the dual-CPU machine; the corresponding LJ speedup (not shown) is 9x. (These results are unrelated to measurements that use multiple networked nodes where near-linear scaling is readily achieved.)

4. Discussion

4.1. Factors influencing GPU performance

Complex GPU architectures often allow a choice of algorithms, some with major impact. The results of the previous section were obtained after investigating different aspects of the algorithms. Although exhaustive testing was not carried out, possible dependencies were examined, some of which are discussed here.

(1) Reordering: For short-ranged interactions, caching can improve performance if atoms are indexed so that data for spatial neighbors are localized in only a few regions of GPU memory.

Table 3. Comparison of parallel MD approaches for the SP fluid based on CPU threads and MPI, with different thread and process counts (n); times per atom-step (t, in μs), speedup relative to serial code, and efficiency relative to the ideal case (no loss due to parallelism) are shown.

	n	t	Speedup	Eff.
1-core		0.103		
CPU	2	0.065	1.58x	79%
	4	0.041	2.51x	63%
	6	0.033	3.12x	52%
	8	0.030	3.43x	43%
MPI	2	0.057	1.81x	91%
	4	0.030	3.43x	86%
	8	0.022	4.68x	59%

This can be accomplished merely by rearranging the storage order of the atoms to correspond to the cells they occupy; atoms in the same cell need not be ordered. The processing speed drops gradually as the interval between reorder operations is increased. Reordering every 100 steps is about optimal, and the speed is 1.15x (SP, $N_e = 96$) slower when the interval is increased to 500 steps. Absent any sorting, the slowdown reaches 4.5x, a major performance loss that exceeds the earlier 2.5x [2]. Note that for measurements that involve the identities of individual atoms, e.g., diffusion, each must be assigned a permanent serial number since storage locations will change.

(2) Thread block size: In the GPU implementation, major loops are replaced by threads that are grouped into blocks. Multiple blocks are processed at the same time, allowing memory access latency to be hidden, but this is limited by model-dependent GPU resources, including any shared memory required by each block and the local (register) storage per thread. For the P4000 there is little sensitivity (3%) to block size over the range 64-1024, so the value 256 was used; this is unlikely to be true in general.

(3) Texture cache: In tests with earlier GPUs the use of the texture cache for reading atom data from global memory produced a substantial (1.8x) speedup. This feature does not influence P4000 performance, although it does affect the somewhat older workstation Tesla K20C, and is likely due to a cache redesign for improving memory access [4].

(4) Periodic boundaries: The extra computation involved is reduced [2] by using 6 bits in the neighbor matrix W entries as flags describing the corrections for each pair ($\pm x$, etc.). Since most of the worked saved is in determining which (if any) corrections are required, 3 bits are sufficient, and the signs determined when needed. This is important in large (but not too large) systems where more than 26 bits of the 32-bit integers are needed for atom indexing.

(5) Neighbor shell width: A thicker shell (δ) entails larger cells and hence more layers and neighbors; the compensation for increased storage is a reduced neighbor refresh rate.The optimal value (for each ρ and T) must be determined empirically; here $\delta = 0.6$ is used. The largest SP(LJ) systems have, on average, 15(48) neighbors, 9(25) layers, and refreshing occurs every 10.2(9.8) time steps. Small irregular speed variations with system size can occur due to the integer number of cells in each grid direction.

(6) Matrix ordering: The access order of the W matrix is a crucial factor. Here, transposing W halves the performance. In the earlier work [2] the performance drop was just 1.1x, suggesting increased sensitivity to memory access issues (the ratio of cores/bandwidth of the GPUs also differs). The very existence of such strong sensitivity serves as a warning that care is required;

even though W is written only when the neighbors are refreshed, it is read every time step, and the matrix arrangement must facilitate coalesced access.

(7) Double-precision performance: On the P4000 the computations ran at about 0.3x the speed, reflecting reduced hardware for d.p. computation. On the K20C the slowdown was just 0.65x. The proportion of work devoted to forces (typically 60%) dominates the d.p. computation, unlike the s.p. case (30%); these comparisons do not reflect nominal GPU GFlop/s rates (d.p. vs s.p.) of 1/32x and 1/3x. Use of d.p. has no obvious effect on the MD results: energy conservation and thermodynamic properties are unchanged; likewise dynamical quantities such as the velocity autocorrelation function of an SP fluid at high density ($\rho = 1.0$, CPU tested), whose oscillations reveal that atoms are caged.

(8) 'Atomic' operations: Their efficiency improves with GPU generation. On the P4000 and the earlier K20C, their use in the cell-layer assignment (N2) had a negligible effect, but with the much older FX770M in [2] 'atomic' operations resulted in 1.12x(1.04x) slowdowns for SP(LJ).

4.2. Alternative approaches

Two other methods are mentioned for comparison purposes.

(1) Cell-block method: An earlier approach to neighbor enumeration [10] associated GPU threads with cells instead of atoms, temporarily storing atom coordinates in shared memory. This was originally compared with the present layer method in [2] using the 32-core FX770M, and later confirmed using a faster 256-core Quadro K4000 GPU (unpublished). The latter measurements showed a layer speedup ranging from 1.2x for LJ with cutoff $r_c = 3.0$, 1.4x for $r_c = 2.5$, to 3.7x for SP, an improvement over the earlier values. Since the principal change to the method is how the neighbor matrix W is constructed, it is interesting to compare the speedups for just this step. The results show corresponding speedup factors of 1.6x, 2.3x, and a substantial 9.2x. Note that the overall time fraction for constructing W increases as r_c is reduced.

(2) All-pairs force evaluation: Even more impressive performance comparisons favoring the GPU can be obtained using a naive MD approach that considers all possible atom pairs. The GPU is able to handle this very efficiently by using shared memory [11], an approach very similar to that recommended for multiplying dense matrices, to increase the amount of computation – albeit mostly unnecessary – performed for each global memory access. The outcome is that the force computations constitute almost the entire workload. The measured speedup of the GPU compared to the CPU is 80x for a relatively small $N_a = 4096$ SP system. Unfortunately the improvement is misleading, and the approach is impractical since the work varies as ${N_a}^2$ rather than N_a; even for this small size it is over 50x slower than the layer method.

4.3. Further MD studies

There are numerous extensions of the basic MD approach for short-range forces that cater to different kinds of systems. Two that require relatively straightforward modifications for the GPU are discussed below; others requiring major changes lie beyond the scope of the present treatment.

(1) Rigid bodies: Modeling molecules based on rigid bodies with multiple interaction sites follows directly from the basic method. A nonspherical structure can be formed, e.g., from four SP spheres in a rigid tetrahedral configuration, spaced to overlap slightly. Spatial orientation can be described in different ways, as quaternions or (as used here) rotation matrices [1]. Integration of the rotational equations of motion employs a generalization of the symplectic leapfrog technique. The additional matrix computations during integration and force/torque evaluation are readily adapted for the GPU, with neighbor processing and forces still dominating the work. This approach was used in a granular segregation study [12]; more complex shapes were employed, together with an SP solvent, in modeling molecular self-assembly [13].

(2) Polymer chains: Polymers are represented by atoms linked with elastic bonds and subject to several interactions. Bond length is governed by the force between bonded atoms; bond angle is regulated by a torque that depends on adjacent bonds; and twist around the bond is (optionally) governed by a dihedral angle that depends on three consecutive bonds. Additionally, there are interactions between nonbonded chain atoms, and (where relevant) between chains and solvent atoms. These interactions can be reformulated for the GPU so that each thread deals with a single atom and all its interactions. In a study of how stiff polymers pack into small shells [14], the length of individual chains (8000 atoms) was insufficient for optimal GPU use. However, since multiple runs from different initial states were required, a set of independent simulations could be carried out simultaneously in a single run – where chains were mutually invisible – to utilize the GPU effectively; the only practical consideration is avoiding excessive cell occupancy, and this was achieved by displacing the shell and chain locations for each realization.

5. Conclusion

The results of this paper show that MD simulation continues to benefit from advances in GPU architecture and performance. These developments, however, do have consequences for developers and users: algorithms become more complex and GPU efficiency remains far below the theoretical peak; these are necessary compromises that do not alter the overall effectiveness of the GPU-based approach.

References

[1] Rapaport D C 2004 *The Art of Molecular Dynamics Simulation* 2nd ed (Cambridge: Cambridge University Press)
[2] Rapaport D C 2011 *Computer Phys. Comm.* **182** 926
[3] *CUDA C Programming Guide* `https://docs.nvidia.com/cuda/pdf/CUDA_C_Programming_Guide.pdf`
[4] *Pascal Tuning Guide* `https://docs.nvidia.com/cuda/pdf/Pascal_Tuning_Guide.pdf`
[5] *NVIDIA Tesla P100 Whitepaper* `http://images.nvidia.com/content/pdf/tesla/whitepaper/pascal-architecture-whitepaper-v1.2.pdf`
[6] Rapaport D C 2006 *Computer Phys. Comm.* **174** 521
[7] Harris M 2007 Optimizing Parallel Reduction in CUDA `https://developer.download.nvidia.com/assets/cuda/files/reduction.pdf`
[8] Harris M, Sengupta S and Owens J D 2007 Parallel Prefix Sum (Scan) with CUDA, *GPU Gems 3*, chap 39 `http://developer.nvidia.com/gpugems3`
[9] *CUDA C Best Practices Guide* `https://docs.nvidia.com/cuda/pdf/CUDA_C_Best_Practices_Guide.pdf`
[10] Anderson J A, Lorenz C D and Travesset A 2008 *J. Comput. Phys.* **227** 5342
[11] Nyland L, Harris M and Prins J 2007 Fast N-Body Simulation with CUDA, *GPU Gems 3*, chap 31 `http://developer.nvidia.com/gpugems3`
[12] Rapaport D C 2014 *J. Phys.: Condens. Matter* **26** 503104
[13] Rapaport D C 2018 *J. Biol. Phys.* **44** 147
[14] Rapaport D C 2016 *Phys. Rev. E* **94** 030401(R)

Thermal transport in periodic and quasiperiodic graphene-hBN superlattice ribbons

Luiz Felipe C Pereira[1,2] and Isaac M Felix[2]

[1] Departamento de Física, Universidade Federal de Pernambuco, Recife 50670-901, Brazil
[2] Departamento de Física, Universidade Federal do Rio Grande do Norte, Natal 59078-970, Brazil

E-mail: pereira@df.ufpe.br

Abstract. Nanostructured superlattices are expected to play a significant role in the next generation of technological devices, specially due to their adjustable physical properties. In terms of heat transport, materials with low thermal conductivities can be useful in thermoelectric devices or heat shields, while materials with high thermal conductivities are fundamental for heat dissipation in miniaturized electronic devices. In general, transport properties are dominated by translational symmetry and the presence of unconventional symmetries might lead to unusual transport characteristics. In this work, we report our results from non-equilibrium molecular dynamics simulations to investigate phonon heat transport in periodic and quasiperiodic graphene-hBN superlattices. The periodic superlattices are built with alternating equal-sized domains of graphene and hBN, while the quasiperiodic case follows the Fibonacci sequence, which lies between periodic and disordered structures. Periodic superlattices can facilitate coherent phonon transport due to constructive interference at the boundaries between the materials. Nonetheless, it is possible to induce a crossover from a coherent to an incoherent transport regime by increasing the length of individual domains, thus adjusting the superlattice period. We also show that the quasiperiodicity can suppress coherent phonon transport in these superlattices. We attribute this behavior to the increased inhomogeneity in the distribution of interfaces, which increases for each Fibonacci generation, hindering coherent phonon transport in the superlattices. The suppression of coherent thermal transport enables a higher degree of control on heat conduction at the nanoscale, and shows potential for application in thermoelectric devices and heat management.

1. Introduction

The term superlattice usually refers to a periodic structure of layers composed of two or more materials. The physical properties of a superlattice can be quite different from the ones of its constituents, originating unusual phenomena. In the case of phonon heat transport, periodic superlattices enable coherent propagation of phonons due to constructive wave interference at the interfaces between the materials. Recent experiments in GaAs/AlAs superlattices [1, 2], epitaxial perovskite oxide superlattices [3] and epitaxial TiN/(Al,Sc)N superlattices [4], have shown that coherent phonons contribute significantly to their heat transport characteristics [5, 6]. Molecular dynamics simulations on superlattices of graphene-hBN [7–9], Si-Ge [10–14], silicene-germanene [15], graphene-nitrogenated holey graphene [16], isotopically modified graphene [17], among others [18–25] also corroborate the presence of coherent phonon transport in periodic superlattices and phononic crystals [26, 27].

In the aforementioned studies, the thermal conductivity of the superlattice is always lower than the conductivity of the pristine materials individually. Increasing the scattering of coherent phonons or preventing their formation is an efficient mechanism to further decrease the lattice thermal conductivity of superlattices, and enable their application in thermal rectification and thermoelectric devices [28–33]. Deviations from periodicity are an effective way of increasing scattering in nanostructures. Indeed, the discovery of quasicrystals spawned a new area in solid state physics and chemistry [34, 35], and one of the first quasiperiodic superlattices artificially produced consisted of a GaAs/AlAs structure following the Fibonacci sequence [36]. Quasiperiodic superlattices are generally formed by two (or more) types of domains, and can be seen as an intermediate between a periodic and a random arrangement [37]. Therefore, quasiperiodic patterning offers a possibility for tuning the physical properties of a given system in a controlled manner [38–40].

Periodic graphene-hBN monolayers have been synthesized via lithography patterning coupled with chemical vapor deposition [41–44] in recent years. This approach enables the fabrication of large-scale hybrid graphene-hBN in-plane heterostructures which are continuous and can be transferred to substrates [44]. Furthermore, these materials have been shown to possess unusual physical properties, quite different from its constituents [41, 45–51]. Indeed, it is reasonable to expect that the thermal conductivity of periodic or quasiperiodic superlattices will be lower than the individual materials because phonons of various wavelengths will be scattered by structures with different length scales.

Here we present our results for the thermal conductivity of periodic and quasiperiodic graphene-hBN superlattice ribbons, obtained from non-equilibrium molecular dynamics simulations. Starting from the periodic superlattice we observe that the thermal conductivity reaches a minimum value as the length of the individual domains increase, which is characteristic of a coherent to incoherent transport transition. Moving on to quasiperiodic ribbons following the Fibonacci sequence, we notice that the domain length at which the conductivity reaches a minimum decreases, and eventually disappears as the Fibonacci generation increases. Since the presence of a valley in the thermal conductivity is associated with a crossover from a coherent to an incoherent transport regime, we conclude that the presence of a quasiperiodicity supresses coherent phonon transport in graphene-hBN superlattices. We attribute this behavior to the increasing number of interfaces per unit cell as the Fibonacci generation increases, which hinders phonon coherence along the superlattice. This suppression of coherent thermal transport in graphene-hBN superlattices enables a further degree of control on heat conduction at the nanoscale, which has the potential to improve the design of novel thermal management devices.

2. Methods

The Fibonacci sequence is an infinite sequence of integer numbers where each element is the sum the two preceding ones. Starting from 0 and 1, the sequence of Fibonacci numbers is: $1, 1, 2, 3, 5, 8, \ldots$. In the present study, we consider a sequence formed by two types of domains A (hBN) and B (graphene), such that each generation is obtained iteratively from its predecessors by an inflation rule A $\rightarrow$ AB and B $\rightarrow$ A. The sequence reads: $B, A, AB, ABA, ABAAB, ABAABABA, \ldots$, as illustrated in Fig. 1(a) up to the fifth generation (5G). The number of domains in each generation grows rapidly and coincides with the respective Fibonacci number. As the generation increases, the ratio between the number of domains of type A and domains of type B approaches the so-called golden ratio: $\frac{N_A}{N_B} \rightarrow (1+\sqrt{5})/2 \approx 1.618$. Therefore, the number of type A domains is always lager than the number of type B domains. Furthermore, domains of type B are never consecutive, whereas two domains of type A can be adjacent to each other.

We build graphene-hBN superlattice nanoribbons by repeating the unit cell of the respective Fibonacci generation, as illustrated in Fig. 1(b). The periodic superlattice case corresponds to

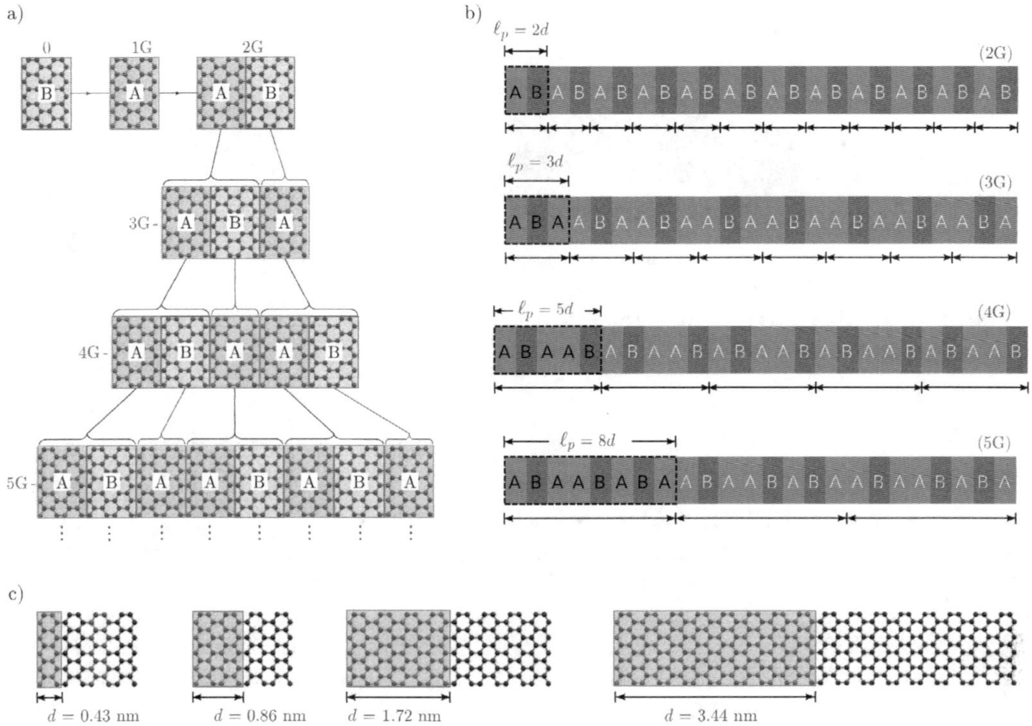

Figure 1. (a) Unit cell of superlattices following the Fibonacci sequence up to fifth generation (5G). Domains of type B correspond to graphene while type A correspond to hBN. (b) Superlattice nanoribbons are obtained by replicating the supercell of each generation. The second generation (2G) is the periodic case. (c) We consider superlattices with equal-sized domains of graphene and hBN, but vary the domain size d. All ribbons have a nominal width of 5 nm.

the second Fibonacci generation (2G). The length of the supercell defines a superlattice period for each generation, which is given by $\ell_p = F_g \cdot d$, where F_g is the Fibonacci number and d is the length of the individual domains. We have considered individual domains of various lengths but constrained the length of graphene domains to equal the length of hBN domains, as shown in Fig. 1(c). Therefore, our samples have equal-sized domains of graphene and hBN for each Fibonacci generation. Finally, all superlattice ribbons considered in this work have a fixed nominal width of 5 nm.

Molecular dynamics simulations were performed with LAMMPS (Large-scale Atomic / Molecular Massively Parallel Simulator) [52]. The interatomic forces were described by the Tersoff empirical potential optimized for carbon and BN nanostructures [53–56], which has been thoroughly tested and employed in several C-BN systems. In all simulations the equations of motion were integrated with a 0.5 fs timestep. The systems were initially thermalized with a Nosé-Hoover thermostat at 300 K for 100 ps. We employed periodic boundary conditions along the length of the superlattice ribbons and free boundary conditions along the other two directions. Each ribbon was relaxed at finite temperature in order to achieve zero-stress along the periodic direction, the stress along the other two directions is automatically zero. The thermostat was turned off after equilibration and the system was then allowed to evolve under microcanonical conditions.

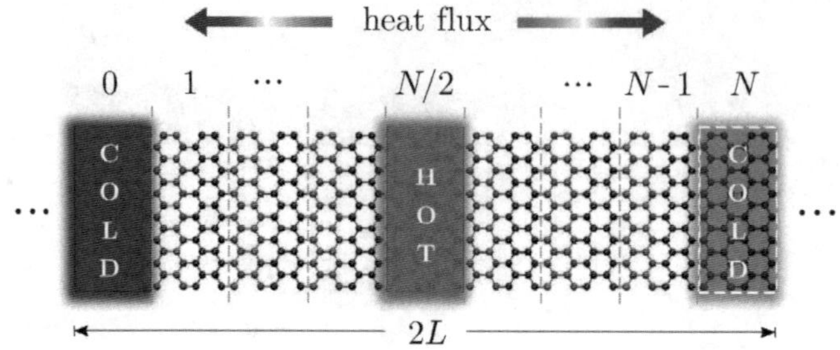

Figure 2. In the RNEMD method the system is divided in N slabs of equal length. Slab 0 is the heat sink and slab $N/2$ is the heat source. Due to periodic boundary conditions, slab N is the image of slab 0.

The lattice (phonon) thermal conductivity was computed with the reverse non-equilibrium molecular dynamics (RNEMD) method [57], which imposes a heat flux through the system by swapping the kinetic energy of atoms in two separate regions, originating a temperature gradient. In the RNEMD method, the system is divided into N slabs along its length and two of those slabs are chosen as heat source and heat sink. Fast moving atoms in the heat sink have their kinetic energy swapped with slow moving atoms in the heat source, generating a temperature gradient. In our setup we choose the first slab as the heat sink and the middle slab as the heat source, as illustrated in Fig. 2. Each slab had and average of 200 atoms.

The heat flux J was calculated from the diference in kinetic energy of the exchanged particles

$$J(t) = \frac{1}{2\Delta t A} \sum_{\text{swaps}} [K_{\text{source}} - K_{\text{sink}}], \tag{1}$$

where Δt is the time interval measured from the beginning of the swaps, K_{source} and K_{sink} are the kinetic energies of the atoms in the heat source and sink, and the sum is over all swaps. We define the cross sectional area A of the nanoribbon as the nominal width of 5 nm multiplied by a thickness of 0.33 nm. The factor of 2 accounts for the fact that the heat flux is divided between both sides of the heat source. When the swaps are introduced the system is taken out of equilibrium, and as they proceed it eventually reaches a steady state where the time average of the heat flux is constant. The frequency with which kinetic energy swaps take place has to be carefully adjusted. Every swap introduces a little discontinuity in the trajectory of the atoms involved. If the swaps happen too often the total energy of the system tends to increase. However, if steps happen too seldom the system takes too long to reach a steady state. In our simulations we performed the energy swaps every 1000 timesteps for a total simulation time of 40×10^7 timesteps, which corresponds to 20 ns.

At every simulation step the temperature of each slab can be calculated from the equipartition theorem as

$$T_i = \frac{2}{3 n_i k_B} \sum_{j=1}^{n_i} \frac{p_j^2}{2 m_j}, \tag{2}$$

where T_i is the temperature of i-th slab, n_i is the number of atoms in it, k_B is Boltzmann's constant, m_j is the mass of atom j, and p_j stands for its linear momentum. From the temperature of each slab it is possible to build a temperature profile for the system, from which the temperature gradient can be measured. Since we are dealing with superlattice nanoribbons, our systems are effectively one-dimensional and only one component of the temperature gradient is relevant. Due to our geometry we have two estimates for the gradient, one from each side of the heat source, as seen in Fig. 2. In fact, one way to estimate if the system has reached the steady state is to check if the gradients on each side of the heat source are equal (within error bars).

Once the system reaches the steady state, both the heat flux and the temperature gradient are stationary and we can calculate the lattice thermal conductivity for a system of length L from Fourier law

$$\kappa(L) = \frac{\langle J \rangle}{\langle \partial T/\partial x \rangle}, \tag{3}$$

where $\langle \cdots \rangle$ represents the time average of the quantities, and x is the direction of the heat flow. In general, the behavior of $\kappa(L)$ can be described by a ballistic-to-diffusive expression such as [58]

$$\kappa(L) = \frac{\kappa}{1 + \Lambda/L} \tag{4}$$

where κ is the intrinsic thermal conductivity of the material, and Λ is an effective phonon mean free path (MFP) [59]. Fan et al. have shown that for pristine graphene sheets the expression above is more accurate when κ and Λ are decomposed between in-plane and out-of-plane components [60]. However, we have verified that for graphene-hBN superlattice ribbons the decomposition is not necessary and Eq. (4) provides accurate results [61]. In what follows we employ Eq. (4) to determine the influence of domain length and Fibonacci generation on the intrinsic thermal conductivity of graphene-hBN nanoribbons.

3. Results and Discussion

We begin by analyzing the dependence of the thermal conductivity with the total length of the periodic superlattice ribbons. At first we consider periodic superlattices with domain sizes $d = 0.43, 0.86, 1.72,$ and 3.43 nm, which correspond to the sizes illustrated in Fig. 1(c). The size dependence of the thermal conductivity for the periodic ribbons is presented in Fig. 3(a), where symbols represent values calculated from Eq. (3), and continuous lines are least-squares fits to the data points using Eq. (4). In order to estimate the uncertainty in the calculated data points we performed independent simulations by changing the initial atomic velocities, and found an uncertainty smaller than 5% in all cases. We have also estimated the uncertainty in κ within an individual simulation by considering the uncertainties in the average heat flux and temperature gradient. In this case we also found an uncertainty smaller than 5%. For the sake of clarity we omit the error bars for superlattices with $L < 100$ nm.

In order to assert the predictive power of the ballistic-to-diffusive expression in Eq. (4), the fitting procedure did not include the three longest systems for each domain size. The error bars represent a 5% uncertainty and are shown only for the largest nanoribbons. The agreement between the fitted lines and the data points even for the largest systems shows the remarkable predictive power of Eq. (4), which can be used to predict the intrinsic lattice thermal conductivity from simulations with relatively short systems. According to Fig. 3(a), simulating nanoribbons of length up to 100 nm is enough to predict the intrinsic thermal conductivity of the graphene-hBN superlattices.

In Fig. 3(b) we plot the intrinsic thermal conductivity as a function of the domain size for periodic superlattices. As the domain size increases from 0.43 to 1.72 nm the thermal conductivity decreases monotonically. Once the domains increase beyond 1.72 nm the conductivity then increases monotonically. The minimum thermal conductivity, $\kappa = 89$ W/m·K at $d = 1.72$ nm, represents a 98% reduction in thermal conductivity when compared to pristine graphene [61–63]. In addition, this valley is a signature of a transition from coherent (wave-like) to incoherent (particle-like) phonon transport, characteristic of superlattices [5, 6, 61, 64]. Therefore, when $d < 1.72$ nm phonons behave mostly as waves and experience constructive

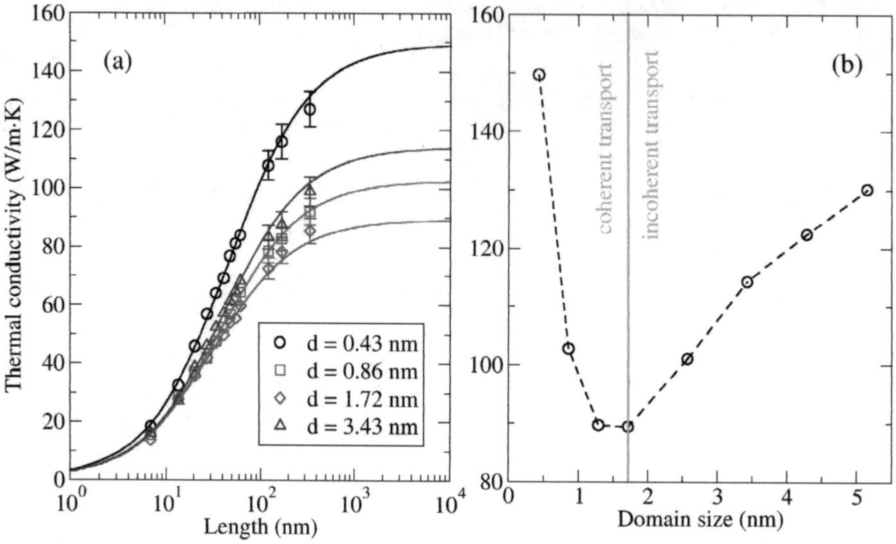

Figure 3. (a) Thermal conductivity versus nanoribbon length for periodic superlattices with various domain sizes. Data points are from RNEMD simulations and lines are least-square fits to the data. (b) Intrinsic thermal conductivity versus domain size for periodic superlattices. The dashed line is a guide to the eye. The minimum thermal conductivity indicates the transition from coherent to incoherent phonon transport.

interference at the interfaces between graphene and hBN domains. Meanwhile, for $d > 1.72$ nm the phonons behave as particles and experience diffuse scattering at the interfaces.

Next we probe the effect of a quasiperiodicity on the thermal conductivity of graphene-hBN superlattice ribbons. In Fig. 4 we present the length dependence of the conductivity from the second Fibonacci generation (2G), i.e. the periodic case, to the eighth generation (8G), with a fixed domain size $d = 0.43$ nm, which corresponds to the smallest domain in Fig. 1(c), including only one hexagon of atoms per domain. We notice a decrease in the intrinsic thermal conductivity as the generation increases. The figure also includes data for a random superlattice with the same domain length. The random lattice was generated by randomly distributing domains of graphene and hBN subject to two constraints: domains of type B (graphene) are never adjacent, and the ratio between A and B domains (N_A/N_B) is as close as possible to the golden ratio. These constraints are necessary in order to compare the quasiperiodic superlattices to equivalent random lattices, since the thermal conductivity of graphene is much larger than hBN, and the superlattice conductivity is strongly dependent on the number of interfaces.

The symbols in Fig. 4 represent values calculated from Eq. (3), while the continuous lines are least-squares fits to the data points using Eq. (4). Again, we performed independent simulations by changing the initial atomic velocities and also considered the uncertainties in the average heat flux and temperature gradient. In both cases the uncertainty in the conductivity was smaller than 5% for all cases. Once again, for the sake of clarity we only present the error bars for the longest superlattices representing a 5% uncertainty. The longest nanoribbons were intentionally excluded from the fitting procedure and assert the predictive power of Eq. (4). The data in Fig. 4 also show that for domains of length $d = 0.43$ nm, the thermal conductivity decreases with the Fibonacci generation, approaching the conductivity of a random superlattice. We attribute this behavior to increased phonon scattering at interfaces, since the number of interfaces per

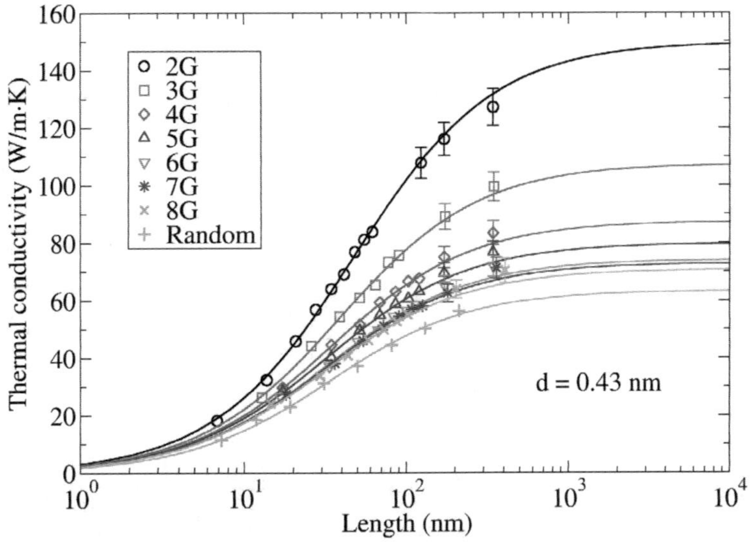

Figure 4. Thermal conductivity as a function of nanoribbon length for quasiperiodic superlattices with $d = 0.43$ nm. Data points are extracted from RNEMD simulations and lines are least-square fits to the data. Notice that κ decreases with the Fibonacci generation and approaches the random superlattice.

unit cell increases with the generation as illustrated in Fig. 1(b). Nonetheless, we notice that even for the eighth Fibonacci generation the thermal conductivity of quasiperiodic superlattices is larger than that of an equivalent random superlattice.

Now, we vary the domain length and analyze its influence on the thermal conductivity. Since we want to focus on the effect of the quasiperiodicity on the thermal transport, we consider only superlattices with equal sized domains of graphene and hBN. Fig. 5 shows the intrinsic thermal conductivity as a function of the domain length from the second to the eighth Fibonacci generation. The figure also includes the conductivity of random superlattices which were built respecting the two constraints in order to make them equivalent to the quasiperiodic ones in terms of adjacent domains and ratio between domain types. A non-monotonic behavior of κ as a function of d, with the presence of a minimum, is a common feature for periodic superlattices, and the presence of this minimum is attributed to the crossover from a coherent transport regime to an incoherent one [5, 6, 61, 64, 65].

In Fig. 5 we notice that the value of d for which κ is minimum decreases as the Fibonacci generation increases. From 7G onwards, including the random superlattice, κ increases monotonically with d and the valley associated with the coherent to incoherent crossover disappears. This observation suggests that the presence of quasiperiodicity gradually suppresses coherent thermal transport in graphene-hBN superlattices, approaching a random superlattice in which the heat transport is predominantly incoherent. The observed suppression is related to the increasing number of interfaces per unit cell, which hinders phonon coherence along the superlattice [65]. Furthermore, the presence of the quasiperiodicity (here from 3G to 8G) decreases the thermal conductivity with respect to the periodic superlattice (2G) with the same domain length.

Finally, we try to quantify the coherent thermal transport in Fibonacci quasiperiodic graphene-hBN superlattices. Following the decomposition of κ in coherent and incoherent

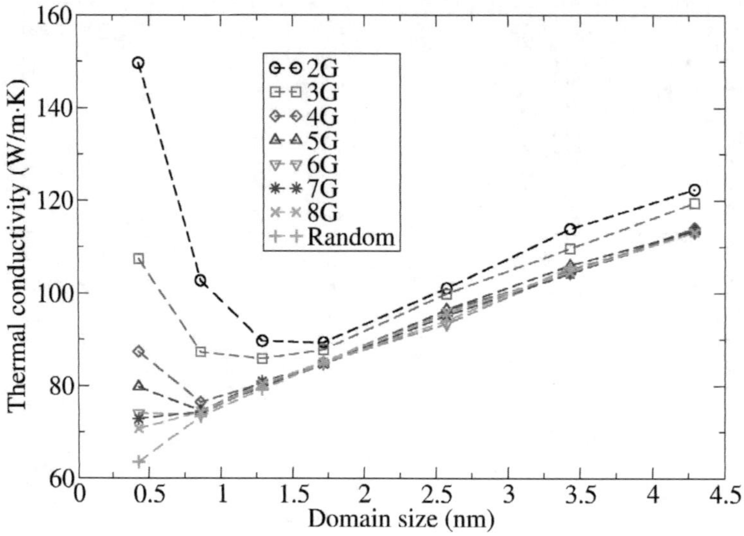

Figure 5. Intrinsic thermal conductivity versus domain size for quasiperiodic superlattices. The minimum conductivity associated with the crossover from coherent to incoherent phonon transport regime disappears for higher generations. Dashed lines are guides to the eye.

phonon contributions proposed by Wang et al. [66], we write

$$\kappa = \kappa_{coherent} + \kappa_{incoherent}. \tag{5}$$

We assume that in the case of periodic and quasiperiodic superlattices both coherent and incoherent phonons contribute to the thermal conductivity, while in random superlattices only incoherent phonons contribute to κ since coherent phonons become localized [26, 27, 66]. Therefore, $\kappa_{incoherent} = \kappa_{random}$, and we obtain the coherent contribution to the thermal conductivity of a quasiperiodic superlattice by subtracting the conductivity of an equivalent random superlattice:

$$\kappa_{coherent} = \kappa - \kappa_{random}. \tag{6}$$

Fig. 6 shows the contribution of coherent transport to the thermal conductivity of quasiperiodic superlattices normalized by their total conductivity. In superlattices with $d = 0.43$ nm the contribution of coherent phonons ranges from 60% in the second generation down to 10% for the eighth generation. For $d = 0.86$ nm only the second and third generations show a contribution larger than 10% from coherent phonons to the total κ. According to Fig. 6, for $d > 1.5$ nm at least 90% of the thermal conductivity in quasiperiodic graphene-hBN superlattices it due to incoherent thermal transport. Therefore, the analysis based on the decomposition of κ further corroborates the conclusions drawn from Fig. 5 regarding the disappearance of the transition from coherent to incoherent heat transport.

We would like to point out that our predictions can be verified in future experimental realizations of graphene-hBN superlattices. The periodic in-plane heterostructures of graphene and hBN produced by Liu et al. [44] presented domain lengths ranging from 100 nm to 1 μm. Due to experimental limitations it is unlikely that superlattices with domain lengths below 10 nm will be produced in the near future. Therefore, according to our simulations the thermal transport in graphene-hBN superlattices experimentally produced should always

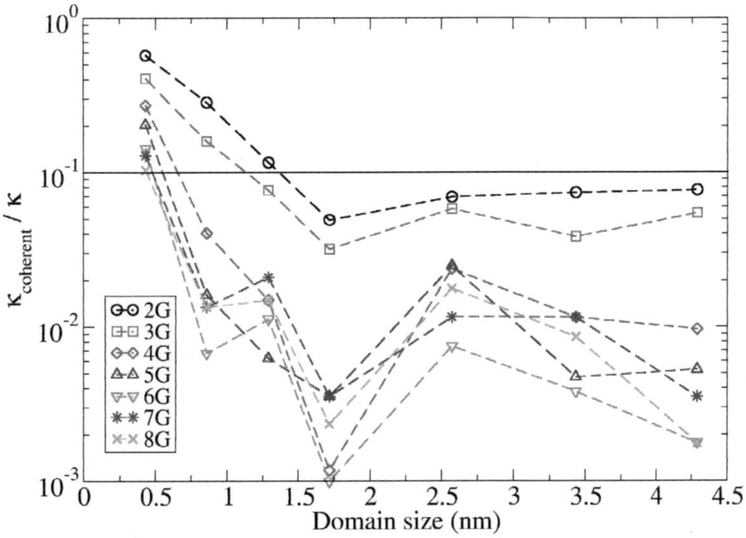

Figure 6. Normalized coherent contribution to the thermal conductivity versus domain length. Below the horizontal line incoherent phonons are responsible for more than 90% of the thermal conductivity.

be in the incoherent regime at room temperature, and their conductivity should increase with the size of the domains [61, 65]. A behavior consistent with these conclusions has recently been observed by experiments in graphene isotope superlattices [33].

4. Conclusions

In summary, we performed large scale non-equilibrium molecular dynamics simulations to systematically investigate the thermal transport properties of periodic and quasiperiodic graphene-hBN nanoribbons. For periodic superlattices we observed a minimum in the thermal conductivity as the length of the individual domains was increased from its smallest value, which is characteristic of a coherent to incoherent transport transition. Moving on to quasiperiodic ribbons following the Fibonacci sequence, we notice that the domain length at which the conductivity reaches a minimum decreases, and eventually disappears as the Fibonacci generation increases. Since the presence of a valley in the thermal conductivity is associated with a crossover from a coherent to an incoherent transport regime, we conclude that the presence of a quasiperiodicity hinders coherent phonon transport in graphene-hBN superlattices. We attribute this behavior to the increasing number of interfaces per unit cell as the Fibonacci generation increases, which supresses phonon coherence along the superlattice. Our prediction could, at least in principle, be measured in experiments at room temperature with current fabrication techniques. The suppression of coherent thermal transport in graphene-hBN superlattices enables a further degree of control on heat conduction at the nanoscale, which has the potential to improve the design of novel thermal management devices.

Acknowledgments

We acknowledge financial support from Coordenação de Aperfeiçoamento de Pessoal de Nível Superior (CAPES), and Conselho Nacional de Desenvolvimento Científico e Tecnológico (CNPq) (Grants 309961/2017 and 436859/2018). Computational resources were provided by the High

Performance Computing Center (NPAD) and the CLIMA cluster at UFRN. LFCP acknowledges a travel grant provided by the University of Georgia, which enabled him to attend the 33rd Annual Center for Simulational Physics Workshop.

References

[1] Luckyanova M N, Garg J, Esfarjani K, Jandl A, Bulsara M T, Schmidt A J, Minnich A J, Chen S, Dresselhaus M S, Ren Z *et al.* 2012 *Science* **338** 936–939
[2] Luckyanova M N, Mendoza J, Lu H, Song B, Huang S, Zhou J, Li M, Dong Y, Zhou H, Garlow J, Wu L, Kirby B J, Grutter A J, Puretzky A A, Zhu Y, Dresselhaus M S, Gossard A and Chen G 2018 *Sci. Adv.* **4** eaat9460
[3] Ravichandran J, Yadav A K, Cheaito R, Rossen P B, Soukiassian A, Suresha S, Duda J C, Foley B M, Lee C H, Zhu Y *et al.* 2014 *Nat. Mater.* **13** 168
[4] Saha B, Koh Y R, Feser J P, Sadasivam S, Fisher T S, Shakouri A and Sands T D 2017 *J. Appl. Phys.* **121** 015109
[5] Simkin M and Mahan G 2000 *Phys. Rev. Lett.* **84** 927
[6] Maldovan M 2015 *Nat. Mater.* **14** 667
[7] Zhu T and Ertekin E 2014 *Phys. Rev. B* **90** 195209
[8] da Silva C, Saiz F, Romero D A and Amon C H 2016 *Phys. Rev. B* **93** 125427
[9] Chen X K, Xie Z X, Zhou W X, Tang L M and Chen K Q 2016 *Appl. Phys. Lett.* **109** 023101
[10] Chen J, Zhang G and Li B 2009 *Appl. Phys. Lett.* **95** 073117
[11] Hu M and Poulikakos D 2012 *Nano Lett.* **12** 5487–5494
[12] Lin K H and Strachan A 2013 *Phys. Rev. B* **87** 115302
[13] Mu X, Wang L, Yang X, Zhang P, To A C and Luo T 2015 *Sci. Rep.* **5** 16697
[14] Kandemir A, Ozden A, Cagin T and Sevik C 2017 *Sci. Technol. Adv. Mater.* **18** 187–196
[15] Wang X, Hong Y, Chan P K L and Zhang J 2017 *Nanotechnol.* **28** 255403
[16] Wang X, Wang M, Hong Y, Wang Z and Zhang J 2017 *Phys. Chem. Chem. Phys.* **19** 24240–24248
[17] Mu X, Zhang T, Go D B and Luo T 2015 *Carbon* **83** 208–216
[18] Chen Y, Li D, Lukes J R, Ni Z and Chen M 2005 *Phys. Rev. B* **72** 174302
[19] Latour B, Volz S and Chalopin Y 2014 *Phys. Rev. B* **90** 014307
[20] Frieling R, Radek M, Eon S, Bracht H and Wolf D 2014 *Appl. Phys. Lett.* **105** 132104
[21] Bracht H, Eon S, Frieling R, Plech A, Issenmann D, Wolf D, Hansen J L, Larsen A N, Ager III J and Haller E 2014 *New J. Phys.* **16** 015021
[22] Xiong S, Latour B, Ni Y, Volz S and Chalopin Y 2015 *Phys. Rev. B* **91** 224307
[23] Mizuno H, Mossa S and Barrat J L 2015 *Sci. Rep.* **5** 14116
[24] Chakraborty P, Cao L and Wang Y 2017 *Sci. Rep.* **7** 8134
[25] Termentzidis K 2018 *J. Phys. D* **51** 094003
[26] Hu S, Zhang Z, Jiang P, Chen J, Volz S, Nomura M and Li B 2018 *J. Phys. Chem. Lett.* **9** 3959–3968
[27] Hu S, Zhang Z, Jiang P, Ren W, Yu C, Shiomi J and Chen J 2019 *Nanoscale* **11** 11839–11846
[28] Liu Y Y, Zhou W X, Tang L M and Chen K Q 2014 *Appl. Phys. Lett.* **105** 203111
[29] Chen X K, Liu J, Xie Z X, Zhang Y, Deng Y X and Chen K Q 2018 *Appl. Phys. Lett.* **113** 121906
[30] Zhang Z, Hu S, Nakayama T, Chen J and Li B 2018 *Carbon* **139** 289–298
[31] Chen X K, Xie Z X, Zhang Y, Deng Y X, Zou T H, Liu J and Chen K Q 2019 *Carbon* **148** 532–539
[32] Zeng Y, Wu D, Cao X, Zhou W, Tang L and Chen K 2020 *Adv. Funct. Mater.* **30** 1903873
[33] Whiteway E, Lee M and Hilke M 2020 *ACS Appl. Nano Mater.* **3** 9167–9173
[34] Shechtman D, Blech I, Gratias D and Cahn J W 1984 *Phys. Rev. Lett.* **53** 1951
[35] Levine D and Steinhardt P J 1984 *Phys. Rev. Lett.* **53** 2477
[36] Merlin R, Bajema K, Clarke R, Juang F Y and Bhattacharya P 1985 *Phys. Rev. Lett.* **55** 1768
[37] Albuquerque E L and Cottam M G 2004 *Polaritons in periodic and quasiperiodic structures* (Elsevier)
[38] Yu Y, Xie Z, Yu X, Li H, Xu P, Yao H and Zhu S 2006 *Phys. Rev. A* **74** 042332
[39] Zhao P L and Chen X 2011 *Appl. Phys. Lett.* **99** 182108
[40] Lang L J, Cai X and Chen S 2012 *Phys. Rev. Lett.* **108** 220401
[41] Ci L, Song L, Jin C, Jariwala D, Wu D, Li Y, Srivastava A, Wang Z, Storr K, Balicas L *et al.* 2010 *Nat. Mater.* **9** 430–435
[42] Liu Z, Song L, Zhao S, Huang J, Ma L, Zhang J, Lou J and Ajayan P M 2011 *Nano Lett.* **11** 2032–2037
[43] Gao Y, Zhang Y, Chen P, Li Y, Liu M, Gao T, Ma D, Chen Y, Cheng Z, Qiu X *et al.* 2013 *Nano Lett.* **13** 3439–3443
[44] Liu Z, Ma L, Shi G, Zhou W, Gong Y, Lei S, Yang X, Zhang J, Yu J, Hackenberg K P *et al.* 2013 *Nat. Nanotechnol.* **8** 119–124

[45] Ding Y, Wang Y and Ni J 2009 *Appl. Phys. Lett.* **95** 123105
[46] Seol G and Guo J 2011 *Appl. Phys. Lett.* **98** 143107
[47] Shinde P P and Kumar V 2011 *Phys. Rev. B* **84** 125401
[48] Kan M, Zhou J, Wang Q, Sun Q and Jena P 2011 *Phys. Rev. B* **84** 205412
[49] Bernardi M, Palummo M and Grossman J C 2012 *Phys. Rev. Lett.* **108** 226805
[50] Jain N, Bansal T, Durcan C A, Xu Y and Yu B 2013 *Carbon* **54** 396–402
[51] Zhou Y, Zhang J, Ye C, Miao X and Zhang D 2014 *J. Appl. Phys.* **115** 114313
[52] Plimpton S 1995 *J. Comput. Phys.* **117** 1–19
[53] Tersoff J 1988 *Phys. Rev. Lett.* **61** 2879
[54] Lindsay L and Broido D A 2010 *Phys. Rev. B* **81** 205441
[55] Sevik C, Kinaci A, Haskins J B and Çagin T 2011 *Phys. Rev. B* **84** 085409
[56] Kinaci A, Haskins J B, Sevik C and Cagin T 2012 *Phys. Rev. B* **86** 115410
[57] Müller-Plathe F 1997 *J. Chem. Phys.* **106** 6082–6085
[58] Schelling P K, Phillpot S R and Keblinski P 2002 *Phys. Rev. B* **65** 144306
[59] Pereira L F C, Mortazavi B, Makaremi M and Rabczuk T 2016 *RSC Adv.* **6** 57773
[60] Fan Z, Pereira L F C, Hirvonen P, Ervasti M M, Elder K R, Donadio D, Ala-Nissila T and Harju A 2017 *Phys. Rev. B* **95** 144309
[61] Felix I M and Pereira L F C 2018 *Sci. Rep.* **8** 2737
[62] Balandin A A, Ghosh S, Bao W, Calizo I, Teweldebrhan D, Miao F and Lau C N 2008 *Nano Lett.* **8** 902–907
[63] Xu X, Pereira L F C, Wang Y, Wu J, Zhang K, Zhao X, Bae S, Tinh Bui C, Xie R, Thong J T L, Hong B H, Loh K P, Donadio D, Li B and Özyilmaz B 2014 *Nat. Commun.* **5** 3689
[64] Hyldgaard P and Mahan G D 1997 *Phys. Rev. B* **56** 10754–10757
[65] Felix I M and Pereira L F C 2020 *Carbon* **160** 335–341
[66] Wang Y, Huang H and Ruan X 2014 *Phys. Rev. B* **90** 165406

AUTHOR INDEX